CODE OF PRACTICE FOR THE CONSTRUCTION, MACHINERY, EQUIPMENT, STABILITY, OPERATION AND EXAMINATION OF MOTOR VESSELS, OF UP TO 24 METRES LOAD LINE LENGTH, IN COMMERCIAL USE AND WHICH DO NOT CARRY CARGO OR MORE THAN 12 PASSENGERS.

SURVEYOR GENERAL'S ORGANISATION
1993

LONDON: HMSO

Cover photographs are reproduced by kind permission of: Capt M H B Snelling & BMIF

First published 1993
ISBN 0 11 551185 7

CODE OF PRACTICE
FOR THE CONSTRUCTION, MACHINERY, EQUIPMENT, STABILITY, OPERATION AND EXAMINATION OF MOTOR VESSELS, OF UP TO 24 METRES LOAD LINE LENGTH, IN COMMERCIAL USE AND WHICH DO NOT CARRY CARGO OR MORE THAN 12 PASSENGERS.

CONTENTS

1 Foreword

1.1 This Code of Practice has a companion Code which deals with sailing vessels. Reference may be made to the companion Code where a sail assisted motor vessel has a significant sailing rig.

1.2 The Code has been developed for application to United Kingdom motor vessels of up to 24 metres Load Line length which are engaged at sea in activities on a commercial basis and which do not carry cargo or more than 12 passengers.

1.3 It should be noted, however, that the class assigned to a United Kingdom vessel is very much dependent upon the terms on which those persons forming the crew have been appointed. It is therefore important to study the contents of Merchant Shipping Notice No. M.1194 which is reproduced in Annex 1.

1.4 The Merchant Shipping (Vessels in Commercial Use for Sport or Pleasure) Regulations 1993 require vessels other than pleasure yachts (or pleasure craft, as the case may be) to comply with the requirements of the Merchant Shipping (Load Line) Rules 1968, SI 1968 No.1053, as amended, or with the requirements of a Code of Practice acceptable to the Secretary of State for Transport as an equivalent standard, unless exempted by the Merchant Shipping (Load Lines) (Exemption) Order 1968, SI 1968 No.1116 as amended. That Order applies to vessels of less than 80 tons register which:-

a) do not carry cargo;
b) do not carry more than 12 passengers;
c) undertake voyages which are neither more than 15 miles (exclusive of any waters designated in the Merchant Shipping (Categorisation of Waters) Regulations 1992 from the point of departure nor more than 3 miles from land.

1.5 The Code is an acceptable Code of Practice for application in accordance with regulation 16 of the Merchant Shipping (Vessels in Commercial Use for Sport or Pleasure) Regulations 1993, SI 1993 No.1072.

1.6 The Code of Practice for the safety of sail training ships, which was the basis of this Code, was developed by the Marine Directorate of the Department of Transport in collaboration with a Working Group following a period of public consultation.

1.7 The organisations involved in the development of the Codes to cover small commercial motor and sailing vessels were as follows:-

American Bureau of Shipping
Association of British Yacht Charter Companies
British Marine Industries Federation
British Ports Association
British Sub-Aqua Club
Bureau Veritas
Det Norske Veritas
Germanischer Lloyd
Lloyd's Register of Shipping
National Federation of Charter Skippers
National Federation of Sea Anglers
National Federation of Sea Schools
Ocean Youth Club Limited
Plymouth City Council
Professional Boatmans Association
Royal Yachting Association
Society of Consulting Marine Engineers and Ship Surveyors
South Hams District Council

South West Ports Association
Sovereign Sailing
Sunsail International Limited
Surveyor General's Organisation, Department of Transport
Yacht Brokers, Designers and Surveyors Association
Yacht Charter Association

1.8 The primary aim in developing the Code has been to set standards of safety and protection for all on board and particularly for those who are trainees or passengers. The level of safety it sets out to achieve is considered to be commensurate with the current expectations of the general public. The Code relates especially to the construction of a vessel, its machinery, equipment and stability and to the correct operation of a vessel so that safety standards are maintained.

1.9 It will be noted that the Code deals with the equally important subject of manning and of the qualifications needed for the senior members of the crew.

1.10 In addition, however, designers and builders of new vessels will need to pay special regard to the intended area of operation and the working conditions to which a vessel will be subjected when selecting the materials and equipment to be used in its construction.

1.11 The builder, repairer or owner/managing agent of a vessel, as appropriate should take all reasonable measures to ensure that a material or appliance fitted in accordance with the requirements of the Code is suitable for the purpose intended having regard to its location in the vessel, the area of operation and the weather conditions which may be encountered.

1.12 The Commission of the European Communities' general mutual recognition clause should be accepted. The clause states:-

Any requirement for goods or materials to comply with a specified standard shall be satisfied by compliance with:-

.1 a relevant standard or code of practice of a national standards body or equivalent body of a Member State of the European Community; or

.2 any relevant international standard recognised for use in any Member State of the European Community; or

.3 a relevant specification acknowledged for use as a standard by a public authority of any Member State of the European Community; or

.4 traditional procedures of manufacture of a Member State of the European Community where these are the subject of a written technical description sufficiently detailed to permit assessment of the goods or materials for the use specified; or

.5 a specification sufficiently detailed to permit assessment for goods or materials of an innovative nature (or subject to innovative processes of manufacture such that they cannot comply with a recognised standard or specification) and which fulfil the purpose provided by the specified standard;

provided that the proposed standard, code of practice, specification or technical description provides, in use, equivalent levels of safety, suitability and fitness for purpose.

1.13 It is important to stress that, whilst all reasonable measures have been taken to develop standards which will result in the production of safe and seaworthy vessels, total safety at sea can never be guaranteed. As a consequence, it is most strongly recommended that the owner/managing agent of a vessel should take out a policy of insurance for all persons who are part of the vessel's complement from time to time. Such insurance should provide cover which is reasonable for claims which may arise. If a policy of insurance is in force, a copy of the certificate of insurance should be either displayed or available for inspection by persons on board the vessel.

1.14 Compliance with the Code in no way obviates the need for vessels and/or skippers to comply with local authority licensing requirements where applicable.

1.15 When a vessel to which the Code is applicable is permanently based abroad and subject to Rules, Regulations and examination by the administration of the country from which it operates, the owner/managing agent may approach a Certifying Authority with the purpose of establishing "equivalence" with the Code.

"Equivalence" should be established for the construction of a vessel, its machinery, equipment, stability, correct operation and examination of the vessel.

The Certifying Authority, when it is satisfied that it is appropriate to do so, may make recommendations for exemption from the Regulations and compliance with the Code in order to issue a certificate based on declaration(s) and report(s) from the administration of the country in which the vessel is permanently based.

The Certifying Authority should make its recommendations to the Department of Transport for approval by the Secretary of State.

1.16 The Organisations listed in 1.7 above were concerned that the ownership of a small commercial vessel by a club should not be seen as a loophole to circumvent the regulations. It is considered that any vessel owned by a proprietary club for use by the members is likely to fall within the scope of the Code.

The Organisations listed in 1.7 above also considered that the officers and committees of members' clubs with responsibility for the maintenance and operation of club owned vessels operated as pleasure yachts could usefully adopt standards set out in the Code as guidelines on safe practice, for the protection of their members.

2 Definitions

In the Code:-

"Accommodation space" means any space, enclosed on all six sides by solid divisions, provided for the use of persons on board;

"Annual examination" means a general or partial examination of the vessel, its machinery, fittings and equipment, as far as can readily be seen, to ascertain that it has been satisfactorily maintained as required by the Code and that the arrangements, fittings and equipment provided are as documented in the Compliance Examination and Declaration report form SCV2;

"Authorised person" means a person who by reason of relevant professional qualifications, practical experience or expertise is authorised by the Certifying Authority chosen by the owner/managing agent from those listed in the Code to carry out examinations required under section 27 of the Code;

"Bare boat charter" means a charter for which the charterer provides the skipper and the crew;

"Category C waters" means waters designated category C waters in the Merchant Shipping (Categorisation of Waters) Regulations 1992, SI 1992 No.2356 and Merchant Shipping Notice No. M.1504;

"Category D waters" means waters designated category D waters in the Merchant Shipping (Categorisation of Waters) Regulations 1992 and Merchant Shipping Notice No. M.1504;

"Certificate" means the certificate appropriate to a vessel to which the Code is applied;

"Certifying Authority" means either the Surveyor General's Organisation of the Department of Transport or one of the organisations authorised by the Department of Transport to:-

(a) appoint persons for the purpose of examining vessels and issuing and signing Declarations of Examinations; and

(b) issue Certificates.

The organisations so authorised by the Department are as follows:-

Lloyd's Register of Shipping
Bureau Veritas
American Bureau of Shipping
Det Norske Veritas
Germanischer Lloyd
The Royal Yachting Association
The Yacht Brokers, Designers and Surveyors Association
The Society of Consulting Marine Engineers and Ship Surveyors;

"Charter" means an agreement between the owner/managing agent and another party which allows that other party to operate the vessel, and the "Charterer" is that other party;

"Code" means this Code;

"Compliance examination" means an examination of the vessel, its machinery, fittings and equipment, by an authorised person, to ascertain that the vessel's structure, machinery, equipment and fittings comply with the requirements of the Code. At least part of the examination should be conducted when the vessel is out of the water;

"Crew" means a person employed or engaged in any capacity on board a vessel on the business of the vessel;

"Efficient" in relation to a fitting, piece of equipment or material means that all reasonable and practicable measures have been taken to ensure that it is suitable for the purpose for which it is intended to be used;

"Existing vessel" means a vessel which is not a new vessel;

"Freeboard" means the distance measured vertically downwards from the lowest point of the upper edge of the weather deck to the waterline in still water;

"Length" means the overall length from the foreside of the foremost fixed permanent structure to the aftside of the aftermost fixed permanent structure of the vessel;

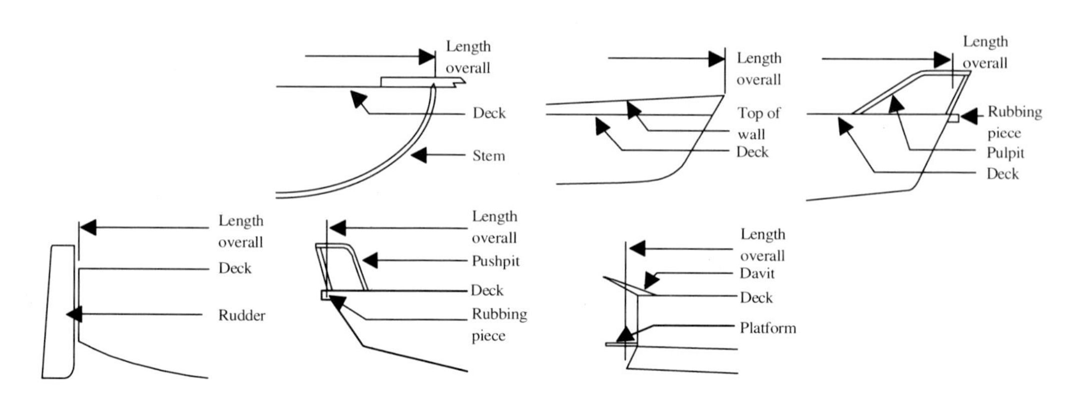

"Load Line length" means either 96% of the total length on a waterline at 85% of the least moulded depth measured from the top of the keel, or the length from the fore side of the stem to the axis of the rudder stock on that waterline, whichever is the greater. In a vessel designed with a rake of keel, the waterline on which this length is measured should be parallel to the design waterline;

"Mile" means a nautical mile of 1852 metres;

"Motor vessel" means a power driven vessel which is not a sailing vessel;

"Multihull vessel" means any vessel which in any normally achievable operating trim or heel angle, has a rigid hull structure which penetrates the surface of the sea over more than one separate or discrete area;

"New vessel" means a vessel to which this Code applies, the keel of which was laid or the construction or lay-up was started on or after 1 April, 1994 or an existing vessel not already being a vessel to which the Code applies but obtained and newly used as such a vessel on or after that date;

"Owner/managing agent" means the registered owner or the owner or managing agent of the registered owner or owner or owner ipso facto, as the case may be, and "Owners/managing agents" should be construed accordingly;

"Passenger" means a person carried in a vessel except:-

(a) a person employed or engaged in any capacity on board the vessel on the business of the vessel;

(b) a person on board the vessel either in pursuance of the obligation laid upon the skipper to carry shipwrecked, distressed or other persons, or by reason of any circumstances that neither the skipper nor the owner nor the charterer (if any) could have prevented; and

(c) a child under one year of age.

Reference should be made to Annex 1 which is Merchant Shipping Notice No. M.1194 - The status of persons carried on United Kingdom ships;

"Person" means a person over the age of one year;

"Pleasure vessel" means a vessel so defined in the Merchant Shipping (Vessels in Commercial Use for Sport or Pleasure) Regulations 1993, SI 1993 No.1072;

"Recess" means an indentation or depression in a deck and which is surrounded by the deck and has no boundary common with the shell of the vessel;

"Safe haven" means a harbour or shelter of any kind which affords entry, subject to prudence in the weather conditions prevailing, and protection from the force of weather;

"Sailing vessel" means a vessel which is designed to be navigated under wind power alone and for which any motor power provided is an auxiliary means of propulsion and/or which possesses a non-dimensional ratio of (sail area) divided by (volume of displacement)$^{2/3}$ of more than 9;

"Skippered charter" means a charter for which the skipper is provided by the owner/managing agent;

SCV1 - means the form for an Application for Examination of a vessel other than an existing vessel which is to be phased-in;

SCV2 - means the report form for a Compliance Examination and Declaration;

SCV3 - means the form for an Application for Phase-in of an existing vessel;

SCV4 - means the form for giving "Notice of Intention" for a vessel which is to be phased-in and for which an Application for Examination has been made on form SCV3;

"To sea" means beyond category D waters, or category C waters if there are no category D waters;

"United Kingdom vessel" means a ship as defined in section 21(2) of the Merchant Shipping Act 1979;

"Watertight" means capable of preventing the passage of water in either direction;

"Weather deck" means the main deck which is exposed to the elements;

"Weathertight" means capable of preventing the admission of a significant quantity of water into the vessel when subjected to a hose test.

3 Application and Interpretation

3.1 Application

3.1.1 Compliance with the Code satisfies the requirements of the Merchant Shipping (Vessels in Commercial Use for Sport or Pleasure) Regulations 1993 to the extent covered by 1.4 above. The Code may be applied to any United Kingdom commercially operated motor vessel of up to 24 metres load line length (and any such motor vessel registered or owned in an other country when it operates from a United Kingdom port) which proceeds to sea, does not carry cargo and does not carry more than twelve passengers.

3.1.2 The Regulations apply to vessels operated by proprietors' clubs and associations and when the owner/managing agent is either corporate or private.

3.1.3 The Code should apply from the 1 April 1994.

3.1.4 The Code applies to monohull and multihull vessels.

3.1.5 It is the responsibility of the owner/managing agent to ensure that a vessel is properly maintained and examined in accordance with the Code.

3.2 Areas of Operation

A vessel may be considered for the issue of a certificate allowing it to operate in one of the following five areas:-

Category 4	-	up to 20 miles from a safe haven, in favourable weather and in daylight;
Category 3	-	up to 20 miles from a safe haven;
Category 2	-	up to 60 miles from a safe haven;
Category 1	-	up to 150 miles from a safe haven;
Category 0	-	unrestricted service.

3.3 Certification

To be issued with a certificate for a particular area of operation a vessel should comply with all the requirements of the Code for that operating area, to the satisfaction of the Certifying Authority.

3.4 Sport Diving, Sea Angling and Other Water Based Activities

The Code deals with safety of the vessel and its occupants but the sport or pleasure activities of those on board are not considered from the particular safety needs which may be relevant to the activities.

The Government's objectives for sport were set out in the document "Sport and Active Recreation" which was published in 1991. The principle of self-determination for sports bodies has been encouraged to the extent that when it has been necessary to impose some form of control on such bodies - such as safety or environmental matters - the policy has usually been to encourage the bodies to adopt voluntary codes or procedures which would have the same effect as regulation.

In 1990, the Minister for Sport commissioned a review into safety in water sports. The review concluded that the current system of self-regulation developed by the governing bodies of sport is sufficient to meet their responsibility for the safety of sports participants.

Vessel owners/managing agents and charterers are recommended to discuss and agree their respective responsibilities for safety before the vessel goes to sea.

3.5 Interpretation

Where a question of interpretation of a part of the Code arises a decision may be obtained on written application to the Deputy Surveyor General 1 in the Surveyor General's Organisation of the Department of Transport, who may consult with others as deemed appropriate.

3.6 Updating of the Code

In addition to the arrangements for interpretation in 3.5 the requirements of the Code will be reconsidered by a standing committee, comprising representatives from the organisations named in 1.7 above, not more than two years after completion of the phase-in period, to take account of experience gained from its application.

Thereafter, the Code will be reviewed by the standing committee at intervals not exceeding five years to take into account experience and any new statutory requirements which apply to other vessels of a similar size or type and, which it might be considered reasonable to apply to vessels operating under the Code.

When new standards are developed and finalised by the EC or ISO and which impact upon the requirements of the Code, amendment of the Code may be considered immediately.

4 Construction and Structural Strength

4.1 General Requirements

4.1.1 A vessel for which the area of operation is more than 20 miles from a safe haven should normally be fitted with a watertight weather deck over the length of the vessel and be of adequate structural strength to withstand the sea and weather conditions likely to be encountered in the intended area of operation.

4.1.2 A vessel which is not fitted with a watertight weather deck in accordance with 4.1.1 should normally be restricted to area category 3 or 4 (up to 20 miles from a safe haven).

4.1.3 A vessel which is an open boat should be restricted to area category 4 and be provided with adequate reserves of buoyancy and stability for the vessel with its full complement of persons to survive the consequences of swamping.

4.2 Structural Strength

4.2.1 General

The design of hull structure and construction should provide strength and service life for the safe operation of a vessel, at its service draught and maximum service speed, to withstand the sea and weather conditions likely to be encountered in the intended area of operation.

4.2.2 Construction materials

4.2.2.1 A vessel may be constructed of wood, fibre reinforced plastic (FRP), aluminium alloy or steel or combinations of such materials.

4.2.2.2 Proposals to use any other material should be submitted to the Certifying Authority for consideration and approval.

4.2.2.3 Requirements for materials used for the construction of inflatable and rigid inflatable boats are given in 4.5.2.

4.2.3 New vessels

4.2.3.1 The hull of a new vessel which has been surveyed and certificated by an United Kingdom Load Line Assigning Authority should be acceptable, subject to presentation of a certificate of construction.

4.2.3.2 United Kingdom Load Line Assigning Authorities are Lloyd's Register of Shipping, the British Technical Committee of American Bureau of Shipping and the British Committees of Bureau Veritas, Det Norske Veritas and Germanischer Lloyd.

4.2.3.3 A new vessel which has not been built under the survey of an United Kingdom Load Line Assigning Authority will be considered to be of adequate strength after a satisfactory examination by an authorised person and if it has been built:-

.1 in accordance with the hull certification standards for small vessels craft, set by one of the Authorities; or

.2 in general accord with the standard of a motor vessel which has a record of at least five years' history of safe operation in an area where the sea and weather conditions are no less severe than those likely to be encountered in the intended area of operation.

4.2.3.4 A new vessel not built in accordance with either 4.2.3.1 or 4.2.3.3 may be specially considered, provided that full information (including calculations, drawings, details of materials and construction) is presented to and approved by the Certifying Authority.

4.2.4 Existing vessels

An existing vessel will be considered to be of acceptable strength if it is in a good state of repair and is:-

4.2.4.1 built to one of the standards described in 4.2.3, for new vessels; or

4.2.4.2 of a design with a record of at least five years' history of safe operation in an area where the sea and weather conditions are no less severe than those likely to be encountered in the intended area of operation.

4.3 Decks

4.3.1 Weather deck

4.3.1.1 A watertight weather deck referred to in 4.1.1 should extend from stem to stern and have positive freeboard throughout, in any condition of loading of the vessel. (Minimum requirements for freeboard are given in section 12.)

4.3.1.2 A weather deck may be stepped, recessed or raised provided the stepped, recessed or raised portion is of watertight construction.

4.3.2 Recesses

(For water freeing arrangements generally, see section 6.)

4.3.2.1 A recess in the weather deck should be of watertight construction and have means of drainage capable of efficient operation when the vessel is heeled to 10 degrees, such drainage to have an effective area, excluding grills and baffles, of at least 20cm^2 for each cubic metre of volume of recess below the weather deck.

4.3.2.2 Alternative arrangements for drainage of a recess may be accepted provided it can be demonstrated that, with the vessel upright and at its deepest draught, the recess drains from a swamped condition within 3 minutes.

4.3.2.3 If a recess is provided with a locker which gives direct access to the interior of the hull, the locker should be fitted with weathertight cover(s). In addition the cover(s) to the locker should be permanently attached to the vessel's structure and fitted with efficient locking devices to secure the cover(s) in the closed position.

4.4 Watertight Bulkheads and Damage Survival

(vessels of 15 metres in length and over or carrying 15 or more persons or operating in area category 0 or 1)

4.4.1 New monohull vessels

When a new monohull vessel is 15 metres in length and over or intended to carry 15 or more persons or operate in area category 0 or 1, watertight bulkheads should be fitted in accordance with the following requirements, except that consideration will be given to the continued acceptance of an existing design which does not meet the requirements in full but is part of a building programme in progress at the time when the Code comes into force for new vessels:-

4.4.1.1 Watertight bulkheads should be so arranged that minor hull damage which results in the free-flooding of any one compartment, will not cause the vessel to float at a waterline which is less than 75mm below the weather deck at any point. Minor damage should be assumed to occur anywhere in the length of the vessel but not on a watertight bulkhead. Standard permeabilities should be used in this assessment as follows :-

Space	Permeability %
Appropriated for stores	60
Appropriated for stores but not by a substantial quantity thereof	95
Appropriated for accommodation	95
Appropriated for machinery	85
Appropriated for liquids	0 or 95 whichever results in the more onerous requirement

4.4.1.2 In the damaged condition, the residual stability should be such that the angle of equilibrium does not exceed 7 degrees from the upright, the resulting righting lever curve has a range to the downflooding angle of at least 15 degrees beyond the angle of equilibrium, the maximum righting lever within that range is not less than 100mm and the area under the curve is not less than 0.015 metre radians.

4.4.1.3 The strength of a watertight bulkhead should be adequate for the intended purpose and to the satisfaction of the Certifying Authority.

4.4.1.4 When pipes, cables, etc penetrate watertight bulkheads, they should be provided with valves and/or watertight glands as appropriate.

4.4.1.5 A doorway fitted in watertight bulkhead should be of watertight construction and be kept closed at sea, unless opened at the discretion of the skipper.

4.4.2 New multihull vessels

Generally, the requirements of 4.4.1 for a new monohull vessel should apply to a new multihull vessel of 15 metres in length and over or intended to carry 15 or more persons or operate in area category 0 or 1.

If a multihull vessel does not meet the damage criteria given in 4.4.1.1 and 4.1.1.2 the results of the calculations should be submitted to the Department of Transport for assessment.

4.4.3 Existing vessels

In the case of an existing vessel which is of 15 metres in length and over or intended to carry 15 or more persons or operate in area category 0 or 1, it is most strongly recommended that modifications, which cause the vessel to meet the standard given by 4.4.1 for a monohull or 4.4.2 for a multihull, be implemented when the vessel undergoes major structural alterations.

4.5 Inflatable Boats

The following requirements should apply to an inflatable or rigid inflatable boat which is proposed for operation under the Code, other than a tender (dinghy) covered by 24:-

4.5.1 General

4.5.1.1 Generally, an inflatable boat or rigid inflatable boat which is intended to operate as an independent vessel under the Code (and is not a tender operating from a vessel) should be of a design and construction which would meet the requirements of chapter III of the 1974 SOLAS Convention, as amended, and the parts of the Annex to IMO Resolution A.689(17) - Testing of Life-Saving Appliances - which are appropriate to the type of boat and subject to the variations which are given in the Code.

4.5.1.2 When production of boats is covered by an approved quality system and boats are built in batches to a standard design, prototype tests on one boat may be accepted for a boat of the same design submitted for compliance with the Code.

4.5.1.3 A boat should be of strength to withstand the sea and weather conditions likely to be encountered in the intended area of operation.

4.5.1.4 An approved boat may be accepted for area category 4 (up to 20 miles from a safe haven in favourable weather and in daylight).

4.5.1.5 A rigid inflatable boat with a substantial enclosure for the protection of persons on board and purpose designed may be considered for operations in area categories 2 or 3, subject to approval by the Certifying Authority.

4.5.2 Construction materials

Materials should satisfy the requirements of chapter III of the 1974 SOLAS Convention, as amended, except that fire-retarding characteristics are not required for the hull material.

4.5.3 New inflatable Boats

4.5.3.1 A new inflatable boat or rigid inflatable boat should satisfy the requirements of chapter III of the 1974 SOLAS Convention, as amended, and be tested in accordance with the requirements of IMO Resolution A.689(17) as appropriate to the intended use of the boat.

As a minimum, tests to verify aspects of strength of structure should include drop and towing. When lifting arrangements are provided in a boat, a lifting (overload) test should be carried out at ambient temperature with the boat loaded with the mass of the full complement of persons and equipment for which it is to be approved. After each test, the boat should not show any signs of damage.

4.5.3.2 A new boat of a type certified as a rescue boat under Merchant Shipping Regulations or provided with a letter of compliance for use as a fast rescue boat for offshore stand-by vessels, or any equivalent certification or compliance, should be accepted as complying with the construction requirements of the Code.

4.5.3.3 A new boat which is not built in accordance with either 4.5.3.1 or 4.5.3.2 may be specially considered, provided that full information (including calculations, drawings, details of materials and construction) is presented to and approved by the Certifying Authority.

4.5.3.4 A permanent shelter provided for the protection of persons on board should be of construction adequate for the intended purpose and the intended area of operation.

4.5.4 Existing inflatable boats

An existing inflatable boat or rigid inflatable boat will be considered to be of acceptable structural strength if it is in a good state of repair and is:-

4.5.4.1 built to one of the standards described in 4.5.3, for a new boat; or

4.5.4.2 of a design with a record of at least five years' history of safe operation in an area where the sea and weather conditions are no less severe than those likely to be encountered in the intended area of operation.

5 Weathertight Integrity

A vessel should be designed and constructed in a manner which will prevent the ready ingress of sea water and in particular comply with the following requirements:-

5.1 Hatchways and Hatches

5.1.1 General requirements

5.1.1.1 A hatchway which gives access to spaces below the weather deck should be of efficient construction and be provided with efficient means of weathertight closure.

5.1.1.2 A cover to a hatchway should be hinged, sliding, or permanently secured by other equivalent means to the structure of the vessel and be provided with sufficient locking devices to enable it to be positively secured in the closed position.

5.1.1.3 A hatchway with a hinged cover which is located in the forward portion of the vessel should normally have the hinges fitted to the forward side of the hatch, as protection of the opening from boarding seas.

5.1.2 Hatchways which are open at sea

In general, hatches should be kept closed at sea. However, a hatch (other than one referred to in 5.2.2) which is to be open at sea for lengthy periods should be:-

.1 kept as small as practicable, but never more than $1m^2$ in plane area at the top of the coaming;

.2 located on the centre line of the vessel or as close thereto as practicable;

.3 fitted such that the access opening is at least 300mm above the top of the adjacent weather deck at side.

5.2 Doorways and Companionways

5.2.1 Doorways located above the weather deck

5.2.1.1 A doorway located above the weather deck which gives access to spaces below should be provided with a weathertight door. The door should be of efficient construction, permanently attached to the bulkhead, not open inwards, and sized such that the door overlaps the clear opening on all sides, and has efficient means of closure which can be operated from either side.

5.2.1.2 A doorway should be located as close as practicable to the centre line of the vessel. However, if hinged and located in the side of a house, the door should be hinged on the forward edge.

5.2.1.3 A doorway which is either forward or side facing should be provided with a coaming the top of which is at least 300mm above the weather deck. A coaming may be portable provided it is permanently secured to the structure of the vessel and can be locked in position.

5.2.2 Companion hatch openings

5.2.2.1 A companion hatch opening from a cockpit or recess which gives access to spaces below the weather deck should be fitted with a coaming, the top of which is at least 300mm above the sole of the cockpit or recess.

5.2.2.2 When washboards are used to close a vertical opening they should be so arranged and fitted that they will not become dislodged in any event.

5.2.2.3 The maximum breadth of the opening of a companion hatch should not exceed 1 metre.

5.3 Skylights

5.3.1 A skylight should be of efficient weathertight construction and should be located on the centre line of the vessel, or as near thereto as practicable, unless it is required to provide a means of escape from a compartment below deck.

5.3.2 When a skylight is an opening type it should be provided with efficient means whereby it can be secured in the closed position.

5.3.3 In a new vessel, a skylight which is provided as a means of escape should be openable from either side.

5.3.4 Unless the glazing material and its method of fixing in the frame is equivalent in strength to that required for the structure in which it is fitted, a portable "blank" should be provided which can be efficiently secured in place in event of breakage of the glazing.

5.4 Portlights

5.4.1 A portlight to a space below the weather deck or in a step, recess, raised deck structure, deckhouse or superstructure protecting openings leading below the weather deck should be of efficient construction.

5.4.2 In a new vessel, a portlight should not be fitted in the main hull below the weather deck, unless the glazing material and its method of fixing in the frame are equivalent in strength to that required for the structure in which it is fitted.

5.4.3 In a new vessel, an opening portlight should not be provided to a space situated below the weather deck.

5.4.4 In an existing vessel, a portlight fitted below the weather deck and not provided with an attached deadlight should be provided with a "blank" (at the rate of 50% for each size of portlight in the vessel), which can be efficiently secured in place in the event of breakage of the portlight.

5.4.4.1 Such a "blank" is not required for a non-opening portlight which satisfies 5.4.2.

5.4.5 An opening portlight should not exceed 250mm in diameter or equivalent area.

5.5 Windows

5.5.1 In a vessel when a window is fitted in the main hull below the weather deck it should provide watertight integrity (and be of strength compatible with size) for the intended area of operation of the vessel.

5.5.2 In a new vessel, a window should not be fitted in the main hull below the weather deck, unless the glazing material and its method of fixing in the frame are equivalent in strength to that required for the structure in which it is fitted.

5.5.3 A window fitted to a space above the weather deck or in the side of a cockpit or recess should be of efficient weathertight construction.

5.5.4 In a vessel which operates more than 60 miles from a safe haven (area category 0 or 1), portable "blanks" should be provided (at the rate of 50% for each size of window), which can be efficiently secured in place in the event of breakage of a window.

5.5.4.1 Such a "blank" is not required for a window which meets the requirements of 5.5.2.

5.6 Ventilators and Exhausts

5.6.1 A ventilator should be of efficient construction and be provided with a permanently attached means of weathertight closure.

5.6.2 A ventilator should be kept as far inboard as practicable and the height above the deck of the ventilator opening should be sufficient to prevent the ready admission of water when the vessel is heeled.

5.6.3 A ventilator which must be kept open, e.g. for the supply of air to machinery or for the discharge of noxious or flammable gases, should be specially considered with respect to its location and height above deck having regard to 5.6.2 and the downflooding angle.

5.6.4 An engine exhaust outlet which penetrates the hull below the weather deck should be provided with means to prevent backflooding into the hull through the exhaust system. The means may be provided by system design and/or arrangement, built-in valve or a portable fitting which can be applied readily in an emergency.

5.7 Air Pipes

5.7.1 When located on the weather deck, an air pipe should be kept as far inboard as possible and have a height above deck sufficient to prevent inadvertent flooding when the vessel is heeled.

5.7.2 An air pipe, of greater than 10mm inside diameter, serving a fuel or other tank should be provided with a permanently attached means of weathertight closure.

5.8 Sea Inlets and Discharges

5.8.1 An opening below the weather deck should be provided with an efficient means of closure.

5.8.2 When an opening is for the purpose of an inlet or discharge below the waterline it should be fitted with a seacock, valve or other effective means of closure which is readily accessible in an emergency.

5.8.3 When an opening is for a log or other sensor which is capable of being withdrawn it should be fitted in an efficient watertight manner and provided with an effective means of closure when such a fitting is removed.

5.8.4 Inlet and discharge pipes from water closets should be provided with shell fittings as required by 5.8.2. When the rim of a toilet is either below or less than 300mm above the deepest waterline of the vessel, anti-syphon measures should be provided.

5.9 Materials for Valves and Associated Piping

5.9.1 A valve or similar fitting attached to the side of the vessel below the waterline, within an engine space or other high fire risk area, should be normally of steel, bronze, copper or other equivalent material.

5.9.2 When unprotected plastic piping is used it should be of good quality and of a type suitable for the intended purpose. If fitted within an engine space or fire risk area, a means should be provided to stop the ingress of water in the event of the pipe being damaged.

6 Water Freeing Arrangements

6.1 When a deck is fitted with bulwarks such that shipped water may be trapped behind them, the bulwarks should be provided with efficient freeing ports.

6.2 The area of freeing ports should be at least 4% of the bulwark area and be situated in the lower third of the bulwark height, as close to the deck as practicable.

6.2.1 A vessel of less than 12 metres in length, having a well deck aft which is fitted with bulwarks all round and which is intended to operate only in favourable weather and no more than 60 miles from a safe haven (area category 2, 3 or 4), should be provided with freeing ports required by 6.2 or may be provided with a minimum of two ports fitted (one port and one starboard) in the transom, each having a clear area of at least 225 sq.cm.

6.3 When a non-return shutter or flap is fitted to a freeing port it should have sufficient clearance to prevent jamming and any hinges should have pins or bearings of non-corrodible material.

6.4 When a vessel has only small side deck areas in which water can be trapped a smaller freeing port area may be accepted. The reduced area should be based on the volume of water which is likely to become trapped.

6.5 In a vessel when freeing ports can not be fitted, other efficient means of clearing trapped water from the vessel should be provided to the satisfaction of the Certifying Authority.

6.6 Structures and spaces considered to be non-weathertight should be provided with efficient drainage arrangements.

7. Machinery

7.1 General Requirement

7.1.1 Generally, machinery installations should comply with the requirements given below. Other installations proposed may be specially considered, provided that full information is presented to and approved by the Certifying Authority.

7.1.2 In the particular case of a proposal to install an inboard petrol engine in a new vessel, full information should be presented to the Department of Transport for approval.

7.2 Diesel engines

A vessel fitted with an inboard engine should be provided with a suitable diesel engine and sufficient fuel tankage for its intended area of operation.

7.3 Petrol engines

7.3.1 In a vessel which is fitted with a watertight weather deck, a petrol engine may be accepted provided that the engine is a suitable outboard type and a fuel tank is fitted whereby either the tank or the complete contents can be jettisoned rapidly and safely and when spillage during fuel handling will drain directly overboard.

7.3.2 In a vessel which is an open boat and restricted to operating in area category 4, a petrol engine may be accepted provided that the engine is a suitable outboard type. Fuel should be stored in portable containers which can be jettisoned readily or in a rigid hull vessel or rigid inflatable boat a fixed-in-place inboard tank may be accepted subject to:-

.1 the tank being constructed of steel or stainless steel, with rounded corners and edges for explosion proofing purposes, located in a safe place and installation complying with 7.4;

Note! (a) Explosafe foils should not be used in a steel tank.

(b) The tank should be tested to at least 0.3 bar.

.2 an intrinsically safe detector of hydrocarbon gas being fitted under or adjacent to the tank (located in a safe place) when the possibility of accumulation of hydrocarbon vapours exists;

.3 the opening of the vent pipe from the petrol tank being protected by a flash proof fitting;

.4 electrical arrangements complying with section 8.

7.3.3 In an existing vessel only, an inboard petrol engine may be accepted provided that the engine is located in an efficient enclosed space to which a fixed fire extinguishing system is fitted.

7.3.3.1 Provision should be made to ventilate the engine space thoroughly before the engine is started.

7.3.4 In an existing vessel, a fixed-in-place inboard petrol tank should meet the requirements of 7.3.2.2, 7.3.2.3 and 7.3.2.4.

7.3.5 In an existing vessel, fuel stored in portable tanks or containers should meet the requirements of 7.3.1 or 7.3.2 as appropriate.

7.3.6 In an existing inflatable boat or rigid inflatable boat, a petrol engine installation should meet the requirements of 7.3.2.

7.4 Installation

7.4.1 The machinery, fuel tank(s) and associated piping systems and fittings should be of a design and construction adequate for the service for which they are intended and should be so installed and protected as to reduce to a minimum danger to persons during normal movement about the vessel, due regard being paid to moving parts, hot surfaces and other hazards.

7.4.2 Means should be provided to isolate a source of fuel which may feed a fire in an engine space fire situation. A valve or cock, which is capable of being closed from a position outside the engine space, should be fitted in the fuel feed pipe as close as possible to the fuel tank.

7.4.3 In a fuel supply system to an engine unit, when a flexible section of piping is introduced, connections should be of a screw type or equivalent approved type. Flexible pipes should be fire resistant/metal reinforced or otherwise protected from fire. Materials and fittings should be of a suitable recognised national or international standard.

In the case of an existing vessel fitted with a diesel engine in which the installation of a flexible section of piping does not immediately meet the requirements, the requirements should be met when existing fittings in the fuel supply system are replaced.

7.5 Engine Starting

7.5.1 An engine should be provided with either mechanical or hand starting or electric starting with independent batteries.

7.5.2 When the sole means of starting is by battery, the battery should be in duplicate and connected to the starter motor via a 'change over switch' so that either battery can be used for starting the engine. Charging facilities for the batteries should be available.

7.6 Portable Generators

7.6.1 When a portable generator powered by a petrol engine is provided, the unit should be stored on the weather deck.

7.6.2 A deck locker or protective enclosure for the portable generator should have no opening(s) to an enclosed space within the hull of the vessel and the locker or protective enclosure should be adequately ventilated and drained.

7.6.3 Petrol provided for the engine should be stored in portable containers or tanks and meet the requirements of 7.7.

7.7 Stowage of Petrol

When petrol in portable containers for use in an outboard engine of a tender (dinghy) is carried on board, the containers should be clearly marked and should be stowed on the weather deck where they can readily be jettisoned and where spillage will drain directly overboard. The quantity of petrol and number of portable containers should be kept to a minimum.

(Requirements for the storage of petrol for propulsion engines of a vessel are given in 7.3.)

8 Electrical Arrangements

8.1 Electrical arrangements should be such as to minimise risk of fire and electric shock.

8.2 Particular attention should be paid to the provision of overload and short circuit protection of all circuits, except engine starting circuits supplied from batteries.

8.3 When general lighting within a vessel is provided by a centralised electrical system, an alternative source of lighting should be provided, sufficient to enable persons to make their way to the open deck and to permit work on essential machinery.

8.4 Batteries and battery systems should be provided as indicated in 7.5.1, 7.5.2 and 16.1.5.

8.4.1 Ventilation of a battery storage space to the open air should be provided, to release the accumulation of gas which is emitted from batteries of all types.

9 Steering Gear

9.1 A vessel should be provided with efficient means of steering.

9.2 The control position should be located so that the person conning the vessel has a clear view for the safe navigation of the vessel.

9.3 When a steering gear is fitted with remote control, arrangements should be made for emergency steering in the event of failure of the control. Arrangements may take the form of a tiller to fit the head of the rudder stock.

10 Bilge Pumping

10.1 Vessels of 15 metres in Length and Over or Carrying 15 or More Persons or Operating in Area Category 0 or 1

10.1.1 A vessel should have an efficient bilge pumping system consisting of at least one hand bilge pump and one engine driven or independent power bilge pump, with suction pipes so arranged that any compartment can be drained when the vessel is heeled up to an angle of 10 degrees. Pumps provided should be situated in not less than two separate spaces.

10.1.2 When considered necessary to protect the bilge suction line from obstruction, an efficient strum box should be provided.

10.1.3 Other means of providing efficient bilge pumping may be considered provided that full information is submitted to and approved by the Certifying Authority.

10.2 Vessels of Less than 15 metres in Length and Carrying 14 or Less Persons and Operating in Area Category 2, 3 or 4

10.2.1 A vessel should be provided with at least two bilge pumps, one of which may be power driven.

10.2.2 A bilge pump should be capable of being operated with all hatchways and companionways closed.

10.2.3 When considered necessary to protect a bilge suction line from obstruction, an efficient strum box should be provided.

10.2.4 Other means of providing efficient bilge pumping may be considered provided that full information is submitted to and approved by the Certifying Authority.

10.3 Bilge Alarm

10.3.1 When propulsion machinery is fitted in an enclosed watertight compartment, a bilge level alarm should be fitted.

10.3.2 The alarm should provide an audible warning at the control position.

11 Intact Stability

11.1 New Vessels

11.1.1 General

11.1.1.1 The standard of stability to be achieved by a new vessel should be dependent on its length, maximum number of persons permitted to be carried and intended area of operation.

11.1.1.2 A vessel of 15 metres in length and over or carrying 15 or more persons or operating in area category 0 or 1, is required to be provided with stability information which is approved by the Certifying Authority and kept on board the vessel.

11.1.1.3 A vessel of less than 15 metres in length and carrying 14 or less persons and operating in area category 2, 3 or 4 is subject to a simplified assessment of stability and is not required to be provided with approved stability information.

11.1.1.4 If a vessel of multi-hull type does not meet the stability criteria given below, the calculations should be submitted to the Department of Transport for assessment.

11.1.1.5 A vessel of unusual form and/or arrangement should be specially considered by the Certifying Authority.

11.1.2 New vessels of 15 metres in length and over or carrying 15 or more persons or operating in area category 0 or 1

11.1.2.1 The lightship weight, vertical centre of gravity (KG) and longitudinal centre of gravity (LCG) of a monohull vessel should be determined from the results of an inclining experiment.

11.1.2.2 The lightship particulars of a multihull vessel should be obtained by a weighing to determine the lightship weight and longitudinal centre of gravity (LCG) and either a careful calculation or an inclining in air to determine vertical centre of gravity (KG).

11.1.2.3 The lightship weight should be increased by a margin for growth, which need not exceed 5% of the lightship weight, positioned at the LCG and vertical centre of the weather deck amidships or KG, whichever is the higher.

11.1.2.4 Curves of statical stability (GZ curves) should be produced for:-

Loaded departure, 100% consumables; and
Loaded arrival, 10% consumables.

11.1.2.5 Buoyant structures intended to increase the range of positive stability should not be provided by fixtures to superstructures, masts or rigging.

11.1.2.6 The curves of statical stability for the loaded conditions should meet the following criteria:-

.1 The area under the righting lever curve (GZ curve) should be not less than 0.055 metre-radians up to 30 degrees angle of heel and not less than 0.09 metre-radians up to 40 degrees angle of heel or the angle of downflooding if this angle is less; and

.2 the area under the GZ curve between the angles of heel of 30 and 40 degrees or between 30 degrees and the angle of downflooding if this is less than 40 degrees, should be not less than 0.03 metre-radians.

.3 The righting lever (GZ) should be at least 0.20 metres at an angle of heel equal to or greater than 30 degrees.

.4 The maximum GZ should occur at an angle of heel of not less than 25 degrees.

.5 After correction for free surface effects, the initial metacentric height (GM) should not be less than 0.35 metres.

11.1.3 New vessels of less than 15 metres in length and carrying 14 or less persons and operating in area category 2, 3 or 4

11.1.3.1 A vessel should be tested in the fully loaded condition (which should correspond to the freeboard assigned) to ascertain the angle of heel and the position of the waterline which results when all persons which the vessel is to be certificated to carry are assembled along one side of the vessel. (The helmsman may be assumed to be at the helm.) Each person may be substituted by a mass of 75kg for the purpose of the test.

The vessel will be judged to have an acceptable standard of stability if the test shows that:-

.1 the angle of heel does not exceed 7 degrees; and

.2 in the case of a vessel with a watertight weather deck extending from stem to stern, as described in 4.1.1, the freeboard to the deck is not less than 75mm at any point.

11.1.3.2 It should be demonstrated by test or by calculation that an open boat, when fully swamped, is capable of supporting its full outfit of equipment, the total number of persons for which it is to be certificated and a mass equivalent to its engine and full tank of fuel.

11.2 Existing Vessels

11.2.1 General

11.2.1.1 The standard of stability required to be achieved by an existing vessel is generally to be as required for a new vessel.

11.2.1.2 A vessel of 15 metres length and over or carrying 15 or more persons or operating in area category 0 or 1 should be provided with approved stability information.

11.2.2 Existing vessels of 15 metres in length and over or carrying 15 or more persons or operating in area category 0 or 1

Unless a vessel is provided with stability information which is approved and relevant to the vessel in its present condition, the vessel should be treated as if it is a new vessel.

11.2.3 Existing vessels of 15 metres in length and over and carrying 14 or less persons and operating in area category 2, 3 or 4

A vessel for which the intended area of operation is not more than 60 miles from a safe haven (area category 2, 3 or 4) should be provided with stability information which is approved and relevant to the vessel in its present condition or may be treated as an existing vessel of less than 15 metres in length.

11.2.4 Existing vessels of less than 15 metres in length and carrying 14 or less persons and operating in area category 2, 3 or 4

11.2.4.1 Generally, a vessel should be treated as if it is a new vessel.

11.2.4.2 It should be demonstrated by test or by calculation that an open boat, when fully swamped, is capable of supporting its full outfit of equipment, the total number of persons for which it is to be certificated and a mass equivalent to its engine and full tank of fuel.

11.2.4.3 When a vessel fails to meet the standards applied to a new vessel a lesser standard may be accepted by the Certifying Authority, provided that the vessel has a record of at least five years' history of safe operation in the intended area of operation.

11.3 **Inflatable Boats** (New and Existing)

The requirements apply to a new or an existing inflatable boat or rigid inflatable boat.

Unless an inflatable boat or rigid inflatable boat is completely in accordance with a standard production type, for which the Certifying Authority is provided with a certificate of approval for the tests which are detailed below, the tests required to be carried out on a boat floating in still water are:-

11.3.1 Stability Tests

The tests should be carried out with the engine and fuel tank fitted or replaced with an equivalent mass. Each person may be substituted by a mass of 75kg for the purpose of the tests:-

11.3.1.1 The number of persons for which an inflatable boat or rigid inflatable boat is to be certified should be crowded to one side, with half this number seated on the buoyancy tube. This procedure should be repeated with the persons seated on the other side and at each end of the inflatable boat or rigid inflatable boat. In each case the freeboard to the top of the buoyancy tube should be recorded. Under these conditions the freeboard should be positive around the entire periphery of the inflatable boat or rigid inflatable boat.

11.3.1.2 Two persons should recover a third person from the water into the inflatable boat or rigid inflatable boat. The third person should feign to be unconscious and be back towards the inflatable boat or rigid inflatable boat so as not to assist the rescuers. Each person involved should wear an approved lifejacket. The stability of the inflatable boat or rigid inflatable boat should remain positive throughout the recovery.

11.3.2 Damage Tests

The tests should be carried out with an inflatable boat or rigid inflatable boat loaded with the number of persons for which it is to be certificated. The engine, fuel tank and full fuel should be fitted, or replaced by an equivalent mass, and all equipment appropriate to the intended use of the inflatable boat or rigid inflatable boat.

The tests will be successful if, for each condition of simulated damage, the persons for which the inflatable boat or rigid inflatable boat is to be certificated are supported within the inflatable boat or rigid inflatable boat. The conditions are:-

.1 with forward buoyancy compartment deflated;

.2 with the entire buoyancy on one side of the inflatable boat or rigid inflatable boat deflated; and

.3 with the entire buoyancy on one side of the bow compartment deflated (where the division of the inflated tube is appropriate to this test).

11.3.3 Swamp Test

11.3.3.1 It should be demonstrated that an inflatable boat or rigid inflatable boat, when fully swamped, is capable of supporting its full outfit of equipment, the total number of persons for which it is to be certificated and a mass equivalent to its engine and full tank of fuel.

11.3.3.2 In the swamped condition the inflatable boat or rigid inflatable boat should not be seriously deformed.

11.3.3.3 The drainage system should be demonstrated at the conclusion of this test.

12 Freeboard and Freeboard Marking

12.1 General

A vessel, other than an inflatable boat or a rigid inflatable boat, should have a freeboard mark placed on each side of the vessel at the position of the longitudinal centre of flotation.

12.2 Minimum Freeboard

The freeboard should be not less than that determined by the following requirements:-

12.2.1 New vessels

A new vessel, other than an inflatable or rigid inflatable boat covered by 12.4, when in still water and loaded with fuel, stores and weights representing the total number of persons certificated to be carried (taken as 75 kg per person) should be upright and:-

.1 in the case of a vessel with a continuous watertight weather deck in accordance with 4.3.1.1, which is neither stepped nor recessed nor raised, have a freeboard measured down from the lowest point of the weather deck of not less than 300mm for a vessel of 7 metres in length or under and not less than 750mm for a vessel of 18 metres in length or over. For a vessel of intermediate length the freeboard should be determined by linear interpolation;

.2 in the case of a vessel with a continuous watertight weather deck in accordance with 4.3.1.2, which may be stepped, recessed or raised, have a freeboard measured down from the lowest point of the well deck of not less than 200mm for a vessel of 7 metres in length or under and not less than 400mm for a vessel of 18 metres in length or over. For a vessel of intermediate length the freeboard should be determined by linear interpolation;

.3 in the case of either an open or partially open vessel, have a clear height of side (i.e. the distance between the waterline and the lowest point of the gunwale*) of not less than 400mm for a vessel 7 metres in length or under and not less than 800mm for a vessel 18 metres in length or over. For a vessel of intermediate length the clear height should be determined by linear interpolation.

* The clear height of the side is to be measured to the top of the gunwale or capping or to the top of the wash strake if one is fitted above the capping.

12.2.2 Existing vessels

.1 Generally, an existing vessel should comply with 12.2.1.

.2 In the case of an existing vessel which is unable to comply with 12.2.1, the Certifying Authority may be prepared to consider a lesser standard of ‘operational freeboard’ or ‘clear height of side’. However, in such a case it will be necessary for the owner/ managing agent to provide the Certifying Authority with a detailed account of the operational history of the vessel. This detailed account should include sea areas normally visited, loaded draught/freeboard/height of side, number of persons usually carried, number of years employed in this mode, together with other details which may be considered relevant.

12.2.3 All vessels

A vessel should be assigned a freeboard which corresponds to the draught of the vessel when fully loaded with fuel, stores and the total number of passengers and crew to be carried (taken as 75 kg per person) plus 25mm, but which in no case should be less than the freeboard required by 12.2.1 or 12.2.2.

12.3 Freeboard Mark and Loading

12.3.1 The freeboard mark referred to above should measure 300mm in length and 25mm in depth. The marking should be permanent and painted black on a light background or in white or yellow on a dark background. The top of the mark should be positioned at the waterline corresponding to the draught referred to in 12.2.3, at the position of the longitudinal centre of flotation, as shown in the sketch below:-

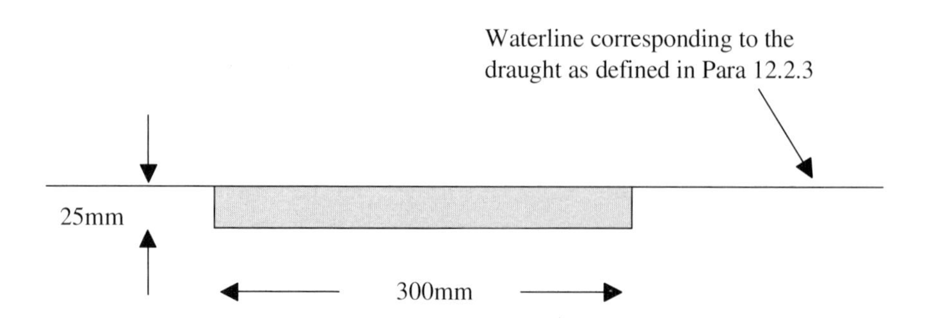

12.3.2 A vessel should not operate in a condition which will result in its freeboard marks being submerged when it is at rest and upright in calm water.

12.4 Inflatable Boats

An inflatable boat or rigid inflatable boat is not required to provided with a freeboard mark.

12.4.1 The freeboard of an inflatable boat or rigid inflatable boat should be not less than 300mm measured from the upper surface of the buoyancy tubes and not less than 250mm at the lowest part of the transom with the inflatable boat or rigid inflatable boat in the following conditions and with the drainage socks (if fitted) tied up:-

.1 the inflatable boat or rigid inflatable boat with all its equipment;

.2 the inflatable boat or rigid inflatable boat with all its equipment, engine and fuel, or replaced by an equivalent mass;

.3 the inflatable boat or rigid inflatable boat with all its equipment and number of persons for which it is to be certificated, having an average mass of 75kg, so arranged that a uniform freeboard is achieved at the side buoyancy tubes; and

.4 the inflatable boat or rigid inflatable boat with all its equipment and number of persons for which it is to be certificated, engine and fuel, or replaced by an equivalent mass, and the inflatable boat retrimmed as necessary to represent a normal operating condition.

13 Life-Saving Appliances

13.1 Life-saving appliances should be provided in accordance with either the requirements appropriate to the type of vessel as given in the Merchant Shipping (Life-Saving Appliances) Regulations 1986, SI 1986 No.1066, as amended, or Annex 2.

13.2 Inflatable liferafts, hydrostatic release units (other than the types which have a date limited life and are test "fired" prior to disposal) and gas inflatable lifejackets should be serviced annually at a service station approved by the manufacturer.

13.3 Orally inflated lifejackets should be pressure tested annually and, as far as is reasonable and practicable, visually examined weekly by the owner/managing agent to determine whether they are safe to use.

13.4 Lifejackets which are not Department of Transport approved should comply with British Standard BS 3595 or equivalent CEN standard and be fitted with a whistle, light and retro-reflective tape.

13.5 Life-saving appliances should be of a type which has been approved by the Department of Transport or which comply with equivalent standards of BSI, CEN or ISO.

14 Fire Safety

14.1 New Vessels

14.1.1 Vessels of 15 metres in length and over

14.1.1.1 In a vessel of 15 metres in length and over, the engine space should be separated from accommodation spaces and store rooms containing combustible materials and liquids.

14.1.1.2 Combustible materials and liquids should not be stowed in the engine space. If non-combustible materials are stowed in the engine space they should be adequately secured against falling into machinery and cause no obstruction to access in or from the space.

14.1.1.3 In a vessel provided with a gas extinguishing system, the boundary of the engine space should be arranged so as to retain the fire extinguishing medium i.e. the engine space should be capable of being closed down in order that the fire extinguishing medium can not penetrate to any other part (or to the outside) of the vessel.

14.1.1.4 Portlights or windows should not be fitted in the boundary of the engine space except that an observation port having a maximum diameter of 150mm may be fitted in an internal boundary bulkhead, provided that the port is of the non-opening type, the frame is constructed of steel or other equivalent material and the port is fitted with a permanently attached cover with securing arrangements. Only fire rated toughened safety glass should be used in an observation port.

14.1.2 Vessels of less than 15 metres in length

14.1.2.1 In a vessel of less than 15 metres in length, the engine should be separated from accommodation spaces by a bulkhead or the engine should be enclosed within a box.

14.1.2.2 Combustible materials and liquids should not be stowed in the engine space. If non-combustible materials are stowed in the engine space they should be adequately secured against falling into machinery and cause no obstruction to access in or from the space.

14.1.2.3 An engine space should be so arranged that, in the event of a fire, the fire extinguishing medium injected can be retained for sufficient time to extinguish the fire.

14.1.3 Insulation

14.1.3.1 Thermal or acoustic insulation fitted inside the engine space should be of non-combustible material.

14.1.3.2 Insulation should be protected against impregnation by flammable vapours and liquids.

14.1.4 Provision for fire extinguishing

Unless a fixed fire extinguishing system is fitted in the engine space, provision should be made in the boundary of the space for discharging fire extinguishing medium into the engine space.

14.1.5 Cleanliness (and pollution prevention)

(Reference should also be made to section 29, Clean Seas.)

14.1.5.1 Provision should be made to retain any oil leakage within the confines of the engine space.

14.1.5.2 In a vessel constructed of wood, measures should be taken to prevent absorption of oil into the structure.

14.1.5.3 In a situation when it is totally impracticable to fit a metal drip tray in way of the engine, the use of the engine bearers as a means of containment of the oil may be accepted when they are of sufficient height and have no limber holes. Provision should be made for the clearing of spillage and drainage collected in the engine space.

14.1.5.4 Efficient means should be provided to ensure that all residues of persistent oils are collected and retained on board for discharge to collection facilities ashore.

14.1.5.5 The engine space should be kept clean and clear of oily waste and combustible materials.

14.1.6 Open flame gas appliances

14.1.6.1 Open flame gas appliances provided for cooking, heating or any other purposes should comply with the requirements of EC Directive 90/396/EEC or equivalent.

14.1.6.2 Installation of an open flame gas appliance should comply with the provisions of Annex 3.

14.1.6.3 Materials which are in the vicinity of open flame cooking or heating appliances should be non-combustible, except that these materials may be faced with any surface finish having a Class 1 surface spread of flame rating when tested in accordance with BS 476: Part 7: 1971 (or any standard either replacing or equivalent to it).

14.1.6.4 Combustible materials and other surfaces which do not have a Class 1 surface spread of flame rating should not be left unprotected within the following distances of the cooker:-

.1 400mm vertically above the cooker, for horizontal surfaces, when the vessel is upright;

.2 125mm horizontally from the cooker, for vertical surfaces.

14.1.6.5 Curtains or any other suspended textile materials should not be fitted within 600mm of any open flame cooking, heating or other appliance.

14.1.7 Furnishing materials

14.1.7.1 Only Combustion Modified High Resilient (CMHR) foams should be used in upholstered furniture and mattresses.

14.1.7.2 Upholstery fabrics should satisfy the cigarette and butane flame tests of British Standard 5852: Part 1: 1979, or equivalent.

14.1.8 Smoke detection

14.1.8.1 In a vessel carrying 15 or more persons, efficient smoke detectors should be fitted in the engine space(s) and spaces containing open flame cooking and/or heating devices.

14.1.8.2 Efficient smoke detectors may be required in order to comply with 14.1.9.2.

14.1.9 Means of escape

14.1.9.1 Each accommodation space, which is either used for sleeping/rest or is affected by a fire risk situation, should be provided with two means of escape. Only in an exceptional case should one means of escape be accepted. Such a case would be when the single escape is to open air or when the provision of a second means of escape would be detrimental to the overall safety of the vessel.

14.1.9.2 In the exceptional case when a single means of escape is accepted, efficient smoke detectors should be provided as necessary to give early warning of a fire emergency which could cut off the single means of escape from a space.

14.2 Existing Vessels

14.2.1 In an existing vessel, the requirements of 14.1 should be carried out as soon as possible, but not later than the phase-in date for the vessel given in Annex 10.

14.2.2 In an existing vessel, replacement of existing upholstery or mattresses to satisfy 14.1.7 may be delayed until renewal.

15. Fire Appliances

A vessel should be provided with efficient fire fighting equipment in accordance with Annex 4.

16 Radio Equipment

16.1 Radio Installation

16.1.1 A vessel should carry equipment for transmitting and receiving on the VHF Maritime Mobile band and for receiving regular shipping weather forecasts for the area of operation.

16.1.2 When the main aerial is fitted to a mast which is equipped to carry sails, an emergency aerial should be provided.

16.1.3 A vessel, other than one operating within 60 miles of a safe haven (area category 2, 3 or 4), should be provided with a radio installation capable of transmitting and receiving messages to and from a radio communications centre on land.

Having regard to the range limitations of VHF, radio equipment should be provided which has a range capability commensurate with that needed for the intended area of operation.

16.1.4 A vessel operating up to 60 miles from a safe haven (area category 2), but in areas where there is a low density of shipping and radio communications centres on land and when the certainty of good VHF coverage is in doubt, should be provided with a radio installation required by 16.1.3.

16.1.5 When the electrical supply to radio equipment is from a battery, charging facilities, or a duplicate battery of capacity sufficient for the voyage, should be provided. Battery electrical supply to radio equipment should be arranged such that radio communications should not be interrupted.

16.1.6 A card or cards giving a clear summary of the radio-telephone distress, urgency and safety procedures should be displayed in full view of the radiotelephone operating position(s). (Guidance on the format of suitable cards is given in Merchant Shipping Notice No. M.1119.)

16.2 406MHz EPIRBs

Requirements for the carriage of a 406MHz EPIRB are given in Annex 2.

17 Navigation Lights, Shapes and Sound Signals

17.1 A vessel should comply with the requirements of the Merchant Shipping (Distress Signals and Prevention of Collisions) Regulations 1989, SI 1989 No.1798, as amended.

17.2 Sound signalling equipment should comply with the Regulations. A vessel of less than 12 metres in length is not obliged to carry the sound signalling equipment required by the Regulations on the condition that some other means of making an efficient sound signal is provided.

17.3 If it can be demonstrated to the Certifying Authority that, for a particular vessel, full compliance with the Regulations is impracticable, application for an exemption should be made to the Department of Transport.

18 Navigational Equipment

18.1 Magnetic Compass

A vessel should be fitted with an efficient magnetic compass and valid deviation card (updated annually) complying with the following requirements as appropriate:-

.1 In a steel vessel, it should be possible to correct the compass for co-efficients B, C and D and heeling error;

.2 The magnetic compass or a repeater should be fitted with an electric light and so positioned as to be clearly readable by the helmsman at the main steering position;

.3 Means should be provided for taking bearings as nearly as practicable over an arc of the horizon of 360 degrees. (This requirement may be met by the fitting of a pelorus or, in a vessel other than a steel vessel, a hand bearing compass.)

18.2 Other Equipment

A vessel which operates more than 20 miles from land (area category 0, 1 or 2) should be provided with:-

.1 a radio navigation aid appropriate for the area of operation;

.2 an echo sounder; and

.3 a distance measuring log; except that this need not be provided where the navigational aid in 18.2.1 provides reliable distance measurements in the area of operation of the vessel.

19 Miscellaneous Equipment

19.1 Nautical Publications

19.1.1 Vessels of 12 metres in length and over

A vessel of 12 metres or more in length should comply with the requirements of the Merchant Shipping (Carriage of Nautical Publications) Rules 1975, SI 1975 No.700.

19.1.2 Vessels of less than 12 metres in length

A vessel of less than 12 metres in length, to which the requirements of 19.1.1 do not apply, should carry up to date charts and, as appropriate, tide tables, a tidal stream atlas and a list of radio signals appropriate to the intended area of operation, and a copy of the international code of signals. (These items may be contained in a Nautical Almanac.)

19.2 Signalling Lamp

A vessel should be provided with an efficient waterproof electric torch suitable for morse signalling.

19.3 Radar Reflector

A vessel should carry a radar reflector complying with the specification ISO 8729 : 1987 - Shipbuilding - Marine radar reflectors (covered by BS 7380 : 1990) or any approved equivalent specification.

19.4 Measuring Instruments

A vessel should carry a barometer.

19.5 Searchlight

A vessel operating in area category 0, 1, 2 or 3 should be provided with an efficient fixed and/ or portable searchlight suitable for use in manoverboard search and recovery operations.

20 Anchors and Cables

20.1 General

The requirements given in Annex 5 are for a vessel of normal form which may be expected to ride-out storms whilst at anchor and when seabed conditions may not be favourable.

20.2 Anchors

20.2.1 The anchor sizes given in Annex 5 are for high holding power (HHP) types.

20.2.2 When a fisherman type of anchor is provided, the mass given in Annex 5 should be increased by 75% but the diameter of the anchor cable need not be increased.

20.2.3 When a vessel has an unusually high windage, due to high freeboard or large superstructures, the mass of anchor given in Annex 5 should be increased to take account of the increase in wind loading.

The diameter of the anchor cable should be appropriate to the increased mass of anchor.

20.3 Cables

20.3.1 The length of anchor cable attached to an anchor should be appropriate to the area of operation but generally should be not less than 4 x the vessel length overall or 30 metres, whichever is the longer, for each of the main and kedge anchors.

20.3.2 In a vessel of 15 metres in length and over, the anchor cable for the main anchor should be of chain.

20.3.3 In a vessel of less than 15 metres in length, the cable for main anchors and for kedge anchors may be of chain or rope.

20.3.4 When the anchor cable is of rope, there should be not less than 10 metres of chain between the rope and the anchor.

20.4 Towline

A vessel should be provided with a towline of not less than the length and diameter of the kedge anchor cable. The towline may be the warp for the kedge anchor.

20.5 Operational

20.5.1 When an anchor mass is more than 30 kg, a windlass should be provided for handling the anchor.

20.5.2 There should be a strong securing point on the foredeck and a fairlead or roller at the stem head which can be closed over the cable.

20.5.3 Area of Operation Category 0, 1, 2, or 3

.1 A vessel should be provided with at least two anchors (one main and one kedge or two main) and cables, subject to 20.1 and in accordance with the requirements of Annex 5.

.2 Anchors of equivalent holding power may be proposed and provided, subject to approval by the Certifying Authority.

20.5.4 Area of Operation Category 4

A vessel of area category 4 is restricted to operations in favourable weather and daylight. It should be provided with at least two anchors (one main and one kedge or two main), the masses of which may be not less than 90% of the requirements of Annex 5, corresponding cables and subject to approval by the Certifying Authority.

21 Accommodation

21.1 General

21.1.1 Hand holds and grab-rails

There should be sufficient hand holds and grab-rails within the accommodation to allow safe movement around the accommodation when the vessel is in a seaway.

21.1.2 Securement of heavy equipment

21.1.2.1 Heavy items of equipment such as batteries, cooking appliance etc., should be securely fastened in place to prevent movement due to severe motions of the vessel.

21.1.2.2 Stowage lockers containing heavy items should have lids or doors with secure fastening.

21.1.3 Access/escape arrangements

Means of escape from accommodation spaces should satisfy the requirements of 5.3.1, 5.3.3 and 14.1.9.

21.1.4 Ventilation

Effective means of ventilation should be provided to enclosed spaces which may be entered by persons on board.

21.2 Vessels at Sea for more than 24 hours

When a vessel is intended to be at sea for more than 24 hours an adequate standard of accommodation for all on board should be provided. In considering such accommodation, primary concern should be directed towards ensuring the health and safety aspects of persons e.g. the ventilation, lighting, water services, galley services and the access/escape arrangements. In particular the following standards should be observed:-

21.2.1 Ventilation

Mechanical ventilation should be provided to accommodation spaces which are situated completely below the level of the weather deck (excluding any coach roof) on vessels intended to make long international voyages or operate in tropical waters and which carry 9 or more berthed persons below deck. As far as practicable, such ventilation arrangements should be designed to provide at least 6 changes of air per hour when the access openings to the spaces are closed.

21.2.2 Lighting

21.2.2.1 An electric lighting system should be installed which is capable of supplying adequate light to all enclosed accommodation and working spaces.

21.2.2.2 The system should be designed and installed in a manner which will minimise the risk of fire and electric shock.

21.2.3 Water services

21.2.3.1 An adequate supply of fresh drinking water should be provided and piped to convenient positions throughout the accommodation spaces.

21.2.3.2 In addition, an emergency (dedicated reserve) supply of drinking water should be carried at the rate of 2 litres per person on board.

21.2.4 Sleeping accommodation

A bunk or cot should be provided for each person on board and at least 50% of those provided should be fitted with lee boards or lee cloths.

21.2.5 Galley

21.2.5.1 A galley should be fitted with a means for cooking and a sink and have adequate working surface for the preparation of food.

21.2.5.2 When a cooking appliance is gimballed it should be protected by a crash bar or other means to prevent it being tilted when it is free to swing and a strap, portable bar or other means should be provided to allow the cook to be secured in position, with both hands free for working, when the vessel is rolling. Means should be provided to isolate the gimballing mechanism.

21.2.5.3 There should be secure storage for food in the vicinity of the galley.

21.2.6 Toilet facilities

21.2.6.1 Adequate toilet facilities, separated from the rest of the accommodation, should be provided for persons on board.

21.2.6.2 In general, there should be at least one marine type flushing water closet and one wash hand basin for every 12 persons.

21.2.7 Stowage facilities for personal effects

Adequate stowage facilities for clothing and personal effects should be provided for each person on board.

22 Protection of Personnel

22.1 Deckhouses

A deckhouse used for accommodation of persons should be of efficient construction.

22.2 Bulwarks, Guard Rails and Handrails

22.2.1 The perimeter of an exposed deck should be fitted with bulwarks, guard rails or guard wires of sufficient strength and height for the safety of persons on deck.

22.2.2 To protect persons from falling overboard, and when the proper working of the vessel is not impeded and there are persons frequently on the deck, bulwarks or three courses of rails or taut wires should be provided and the bulwark top or top course should be not less than 1000mm above the deck (in accordance with Load Line rules). Intermediate courses should be evenly spaced.

22.2.3 In a vessel fitted with a cockpit which opens aft to the sea, additional guardrails should be fitted so that there is no vertical opening greater than 500mm.

22.2.4 Access stairways, ladderways and passageways should be provided with handrails.

22.2.5 In an inflatable boat or a rigid inflatable boat, handgrips, toeholds and handrails should be provided as necessary to ensure safety of all persons on board during transit and the worst weather conditions likely to be encountered in the intended area of operation.

22.3 Safety Harnesses

22.3.1 A vessel should be provided with 2 safety harnesses.

22.3.2 Efficient means for securing the life lines of safety harnesses should be provided on exposed decks, and grabrails provided on the sides and ends of a deckhouse.

22.3.3 Fastening points for the attachment of safety harness life lines should be arranged having regard to the likely need for work on or above deck. In general, securing points should be provided in the following positions:-

.1 close to a companionway; and

.2 on both sides of a cockpit.

22.3.4 When guard rails or wires are not otherwise provided, jackstays (which may be fixed or portable) secured to strong points, should be provided on each side of the vessel to enable crew members to traverse the length of the weather deck in bad weather.

22.4 Toe Rails

When appropriate to the working of a vessel provided with a sailing rig, a toe rail of not less than 25mm in height should be fitted around the working deck.

22.5 Surface of Working Decks

22.5.1 The surface of a working deck should be non-slip.

22.5.2 Acceptable surfaces are: unpainted wood; a non-skid pattern moulded into FRP; non-slip deck paint; or an efficient non-slip covering.

22.5.3 Particular attention should be paid to the surface finish of a hatch cover when it is fitted on a working deck.

22.5.4 In an inflatable boat or rigid inflatable boat the upper surface of the inflated buoyancy tube should be provided with a non-slip finish.

22.6 Recovery of Persons from the Water

An overside boarding ladder or scrambling net which extends from the weather deck to at least 600mm below the operational waterline or other means to aid the recovery of an unconscious person from the water should be provided to the satisfaction of the Certifying Authority.

22.7 Personal Clothing

It should be the responsibility of an owner/managing agent/skipper to advise that the following requirements for items of personal clothing should be met:-

.1 Each person on board a vessel should have protective clothing appropriate to the prevailing air and sea temperatures.

.2 On a vessel which intends to operate in high latitudes, each person on board should have either an approved immersion suit or a dry suit of suitable quality to reduce the likelihood of hypothermia should the wearer enter the sea.

.3 Each person on board a vessel should have footwear having non-slip soles, to be worn on board.

23 Medical Stores

A vessel should carry medical stores appropriate to the area of operation.

23.1 A vessel operating in area category 2, 3 or 4 or on bareboat charter should carry an augmented first aid kit as detailed in Annex 6.

23.2 A vessel, other than on bareboat charter, operating in area category 1 and carrying 14 or less persons should carry the medical stores prescribed by Scale IIA of The Merchant Shipping (Medical Stores) Regulations 1986, SI 1986 No.144, as amended. (The stores are listed in Merchant Shipping Notice M.1319.)

When 15 or more persons are carried the quantities of consumable items prescribed by Scale IIA should be doubled.

23.3 A vessel, other than on bareboat charter, operating in area category 0 should carry the medical stores required by The Merchant Shipping (Medical Stores) Regulations 1986, SI 1986 No.144, as amended or, the owner/managing agent should prepare a plan for providing medical care for all persons on board and submit it to the Department of Transport for approval, not less than 21 days before the intended sailing date.

In addition to detailing the medical stores to be carried, the application must give full details of the intended area of operation and anticipated duration of the voyage, details of the medical, nursing or first aid qualifications of the person(s) who will be responsible for the medical care and any back up facilities (i.e. other ships in company) that will be available.

24 Tenders (Dinghies)

24.1 An inflatable tender is not required to meet the requirements for inflatable boats or rigid inflatable boats in section 4.5.

24.2 A tender should be clearly marked with the number of people of mass 75 kg that it can safely carry and with the name of the parent vessel.

24.3 An inflatable tender should be fit for the purpose intended, regularly inspected by the owner/ managing agent and maintained in a safe condition.

25 Sailing Vessel Features

When a motor vessel is provided with a sailing rig which causes the vessel to be categorised as a sailing vessel the requirements of the companion Code, The Safety of Small Commercial Sailing Vessels - A Code of Practice, apply.

A "sailing vessel" is a vessel which is designed to be navigated under wind power alone and for which any motor power provided is an auxiliary means of propulsion and/or which possesses a non-dimensional ratio of (sail area) divided by (volume of displacement)$^{2/3}$ of more than 9.

26 Manning

A vessel should be safely manned.

26.1 Vessels Other than those on Bare-boat Charter

26.1.1 The qualification of the skipper (and of the other member(s) of the crew, where applicable) for operations in various areas is the subject of a General Exemption from relevant Regulations.

26.1.2 The conditions applicable to the General Exemption and the responsibility of the owner/ managing agent for the safe manning of a vessel are given in Annex 7.

26.2 Vessels on Bare-boat Charter

26.2.1 A vessel operating on bare-boat charter as a pleasure vessel is not subject to the safe manning conditions given in Annex 7.

26.2.2 The owner/managing agent of a vessel offered for bare-boat charter should ensure that the skipper and crew of the vessel are provided with sufficient information about the vessel and its equipment to enable it to be navigated safely. The owner/managing agent should be satisfied that the bare-boat charter skipper and crew are competent for the intended voyage. Details of handover procedures are given in Annex 8.

26.3 Vessels on Skippered Charter

The skipper of a vessel on skippered charter should ensure that each person on board is briefed on safety in accordance with the requirements given in Annex 9.

27 Compliance Procedures, Certification, Examination and Maintenance

27.1 Definitions

For the purpose of an examination:-

"Authorised person" means a person who by reason of relevant professional qualifications, practical experience or expertise is authorised by the Certifying Authority chosen by the owner/managing agent from those listed in the Code to carry out examinations required under section 27 of the Code.

"Compliance examination" means an examination of the vessel, its machinery, fittings and equipment, by an authorised person, to ascertain that the vessel's structure, machinery, equipment and fittings comply with the requirements of the Code. At least part of the examination should be conducted when the vessel is out of the water;

The Certifying Authority should decide the extent of the examination based on the type, age and history of the vessel.

"Annual examination" means a general or partial examination of the vessel, its machinery, fittings and equipment, as far as can readily be seen, to ascertain that it has been satisfactorily maintained as required by the Code and that the arrangements, fittings and equipment provided are as documented in the Compliance Examination and Declaration report form SCV2.

SCV1 - means the form for an Application for Examination of a vessel other than an existing vessel which is to be phased-in.

SCV2 - means the report form for a Compliance Examination and Declaration.

SCV3 - means the form for an Application for Phase-in of an existing vessel.

SCV4 - means the form for giving "Notice of Intention" for a vessel which is to be phased-in and for which an Application for Examination has been made on form SCV3.

27.2 Requirements for Vessels to be Examined and Certificated

27.2.1 The owner/managing agent of a new vessel, or an existing vessel required by 27.4.10 to be treated as a new vessel, to be operated under the Code, should:-

.1 choose an authorised Certifying Authority and contact them to obtain a copy of their Application for Examination form SCV1;

.2 complete form SCV1 and return it to the Certifying Authority; and

.3 arrange with the Certifying Authority for the vessel to be examined by an authorised person and documented on the report form for a Compliance Examination and Declaration, SCV2, as being in compliance with the Code; and

.4 be in receipt of a Small Commercial Vessel Certificate for the vessel prior to it entering into service. (The form of the certificate is given in Annex 11.)

27.2.2 An existing vessel operating under the phase-in arrangements given in 27.4 should hold a "Notice of Intention" form SCV4, to indicate that the owner/managing agent has requested the chosen Certifying Authority to certificate the vessel.

27.3 Arrangements for Motor Training Vessels

No special requirements apply to a motor vessel used for the purpose of providing instruction in operation, navigation and seamanship of such a vessel.

27.4 Phase-in Arrangements for Existing Vessels

27.4.1 An existing vessel is required to be examined, documented and certificated according to the timescale given in Annex 10.

27.4.2 To operate an existing vessel under the terms of the phase-in arrangements the owner/managing agent should make an application for examination not later than 1 January 1994 by arranging to:-

.1 choose an authorised Certifying Authority and contact them to obtain a copy of their Application for Phase-in form SCV3;

.2 complete form SCV3 and return it to the Certifying Authority to register the intention to have the vessel examined and certificated by the phase-in date given in Annex 10.

27.4.3 When making an application, the owner/managing agent will be required to:-

.1 report to the Certifying Authority, the vessel's name, length and the maximum number of persons to be carried;

.2 submit sufficient stability information to allow the Code requirements for stability (section 11) to be assessed by the Certifying Authority;

.3 declare to the Certifying Authority:-

a) within how many miles of a safe haven the vessel is to be operated, that the structural strength of the vessel is considered to be as required by the Code (4.2.4) and that it is in a good state of repair, that the vessel has previously been in commercial use in similar areas of operation and is considered to be satisfactory for such service;

b) that, after 1 April 1994, the life-saving appliances (section 13 and annex 2), fire fighting appliances (section 15 and annex 4), navigation equipment (section 18), miscellaneous equipment (section 19), anchors and cables (section 20 and annex 5), safety harnesses (section 22.3), and medical stores (section 23 and annex 6) will be in accordance with the Code for the intended area of operation; and

c) that the vessel will be manned at all times in accordance with the Code (section 26 and annex 7) for the intended area and type of operation; and

d) that when approved stability information is required (section 11), it will be carried on board at all times; and

e) that the vessel will be maintained as required by the Code and submitted for examination by the phase-in date given in Annex 10.

27.4.4 On receipt and review of the completed Application for Phase-in form SCV3 and the required stability information, the Certifying Authority, when it is considered appropriate to do so, will acknowledge and accept receipt of the application by issuing a "Notice of Intention" form SCV4 which confirms the intention to have the vessel certificated under the Code, acceptance of the stability information and the phase-in date for examination and certification and gives the vessel a unique identification number.

27.4.5 The "Notice of Intention" should be retained on board the vessel at all times and, when the phase-in period is more than 15 months, the owner/managing agent of a vessel should make an annual declaration, reaffirming 27.4.3.3, not more than 3 months before or more than 3 months after the anniversary date of the initial declaration.

27.4.6 Proposals at any time to change any of the information indicated on the Application for Phase-in form SCV3 and/or confirmed in the "Notice of Intention" form SCV4, should not be implemented until they have been submitted to the Certifying Authority on a new Application for Phase-in form SCV3 and the owner/managing agent is in receipt of a revised "Notice of Intention" form SCV4, issued by the Certifying Authority in accordance with the procedure indicated in 27.4.4.

27.4.7 During the phase-in period, a vessel having been declared by the owner/managing agent to be operating under the Code may be examined by the Certifying Authority to confirm the details given in the declaration made by the owner/managing agent.

27.4.8 The owner/managing agent of a vessel should ensure that the vessel is operated in accordance with the declaration made and should arrange for the necessary compliance examination and documentation of the arrangements, fittings and equipment, by an authorised person to be completed and the small commercial vessel certificate issued by the date agreed with the Certifying Authority.

27.4.9 When a vessel issued with a "Notice of Intention" is sold during the phase-in period, the notice is cancelled automatically and the owner/managing agent selling the vessel should return the notice to the Certifying Authority and advise the name and address of the new owner/ managing agent. When the business of the vessel under the new owner/managing agent is covered by the Code, the new owner/managing agent should make an application to the Certifying Authority to operate the vessel under the terms of the phase-in arrangements, in accordance with 27.4.2, before putting the vessel into service.

27.4.10 An existing vessel for which the owner/managing agent has not made application for phase-in on form SCV3 by 1 January, 1994 should be treated as a new vessel under the Code.

27.4.11 A vessel for which the keel was laid or construction or lay-up was started between 1 January 1994 and 31 March 1994 (inclusive of both dates) may be treated as an existing vessel under the Code provided application for phase-in is made on form SCV3 before 1 April 1994.

27.5 Issue of a Certificate of Compliance Under the Code

27.5.1 The owner/managing agent should arrange for a compliance examination to be carried out by an authorised person. The arrangements, fittings and equipment provided on the vessel are to be documented on the Compliance Examination and Declaration report form SCV2. Upon satisfactory completion and documentation of the compliance examination and the owner/ managing agent's and authorised person's declarations, a copy of the report form SCV2 should be forwarded to the Certifying Authority.

27.5.2 In the case of an existing vessel with a "Notice of Intention" form SCV4, when the compliance examination, documentation and required declarations are completed within the 6 month period preceding the phase-in date for examination and if it is considered appropriate to do so, the authorised person may recommend to the Certifying Authority that the certificate should be issued for up to five years from the phase-in date.

27.5.3 Vessels of 15 metres in length and over or carrying 15 or more persons or operating in area category 0 or 1.

.1 Before a certificate is issued the owner/managing agent should be in possession of an approved stability information booklet for the vessel.

.2 An existing vessel (which is not required to be treated as a new vessel) of 15 metres in length and over but carrying 14 or less persons and operating in area category 2, 3 or 4 should satisfy the stability requirements given in 11.2.3.

27.5.4 Vessels of less than 15 metres in length and carrying 14 or less persons and operating in area category 2, 3 or 4

Before a certificate is issued, the owner/managing agent should provide the Certifying Authority with information necessary to confirm that the stability of the vessel meets the standard required by the Code for the permitted area of operation.

27.5.5 Upon satisfactory review of the documented arrangements, fittings and equipment provided in compliance with the Code, also the required declarations in the completed report form SCV2 and approval as appropriate of either the stability information booklet or required stability information, the Certifying Authority will issue the final full term certificate. (The form of the certificate is given in Annex 11.)

27.5.6 A certificate should be valid for not more than five years.

27.6 Compliance and Annual Examinations

27.6.1 Compliance examination for renewal of a certificate

27.6.1.1 The owner/managing agent should arrange for a compliance examination to be carried out by an authorised person from the chosen Certifying Authority. Upon satisfactory completion and verification that the arrangements, fittings and equipment documented in the Compliance Examination and Declaration report form SCV2 remain in compliance with the Code and that the vessel and its machinery are in a sound and well maintained condition, the certificate in force should be endorsed to indicate a 3 month extension and a copy of the report recommending the renewal of the certificate should sent to the Certifying Authority.

27.6.1.2 Upon satisfactory review of the arrangements, fittings and equipment documented in the report form SCV2 as being in compliance with the Code, the Certifying Authority should renew the vessel's certificate.

27.6.2 Vessels of 15 metres in length and over or carrying 15 or more persons

27.6.2.1 Annual examinations by the Certifying Authority

The owner/managing agent should arrange for an annual examination of a vessel as defined in 27.1 to be carried out by an authorised person once in each calendar year, at intervals not exceeding 15 months. On satisfactory completion of the annual examination, the authorised person should enter details of the examination on the Compliance Examination and Declaration report form SCV2 and report the results of the examination to the Certifying Authority.

27.6.3 Vessels of less than 15 metres in length and carrying 14 or less persons

27.6.3.1 Annual examinations by the owner/managing agent

.1 The owner/managing agent should carry out or arrange for an examination of a vessel, once in each calendar year, at intervals not exceeding 15 months, to confirm that the arrangements, fittings and equipment provided on board are in a satisfactory condition and remain as documented in the Compliance Examination and Declaration report form SCV2. Also that the vessel, its machinery, fittings and equipment are in a sound and well maintained condition.

.2 The owner/managing agent should enter details of a successful examination on the report form SCV2 and report the results of the examination to the Certifying Authority.

.3 The owner/managing agent should not complete details on the report form SCV2 if the examination reveals that either the vessel, its machinery, fitting or equipment are not sound or they do not comply with those documented in the Compliance Examination and Declaration report form SCV2. The reasons for the owner/managing agent not being allowed to enter details of the examination on the report form SCV2 should be reported immediately to the Certifying Authority for action as necessary. Also, see 27.9.2.

27.6.3.2 Other examinations by the Certifying Authority

In addition to the compliance examination carried out on behalf of the Certifying Authority and the periodic examinations required to be undertaken by the owner/managing agent, an examination equivalent to the annual examination defined in 27.1 should be carried out on behalf of the Certifying Authority, by an authorised person at least once during the life of the certificate in order that the interval between successive examinations by an authorised person should not exceed 3 years. The owner/managing agent should arrange with the Certifying Authority for this examination to be carried out. On satisfactory completion of the examination, the authorised person should enter details of the examination on the report form SCV2 and report the results of the examination to the Certifying Authority.

27.7 Appeal Against the Findings of an Examination

If an owner/managing agent is dissatisfied with the findings of an examination and agreement can not be reached with the authorised person who carried out the examination, the owner/managing agent may appeal to the Certifying Authority to review the findings. At this review, the owner/managing agent may call a representative or professional adviser to give opinions in support of the argument against the findings of the examination.

Should the above procedures fail to resolve the disagreement, the owner/managing agent may refer the disagreement to the Deputy Surveyor General 1 in the Surveyor General's Organisation of the Department of Transport for arbitration.

27.8 Maintaining and Operating the Vessel

27.8.1 The Certifying Authority may examine a certificated vessel at any time.

27.8.2 It is the responsibility of the owner/managing agent to ensure that at all times a vessel is maintained and operated in accordance with the requirements of the Code, the arrangements as documented in the Compliance Examination and Declaration report form SCV2 and any conditions stated on the vessel's certificate. If for any reason the vessel does not continue to comply with any of these requirements, the owner/managing agent should notify the Certifying Authority immediately. Also see 27.9.2.

27.8.3 If a vessel suffers a collision, grounding, fire or other event which causes major damage, the owner/managing agent should notify the Certifying Authority immediately.

In addition, the owner/managing agent has a statutory requirement to report accidents. The statutory requirements are given in the Merchant Shipping (Accident Investigation) Regulations 1989, SI 1989 No.1172. Merchant Shipping Notice No. M.1383 explains the Regulations and the requirement to report accidents to the Department of Transport.

27.8.4 The nature and extent of major repairs should be subject to the approval of the Certifying Authority.

27.9 Other Conditions Applying to Certificates

27.9.1 Existing vessels with certificates

When an existing vessel has a current Load Line or Load Line Exemption certificate, it may continue to operate under the conditions applicable to the certificate in force. Upon expiry of the certificate in force, the owner/managing agent may choose to renew the previous certificate or apply to a Certifying Authority (on form SCV1) for issue of a small commercial vessel certificate as an existing vessel under the Code.

27.9.2 Validity and cancellation of certificates

27.9.2.1 The validity of a certificate issued under the Code is dependent upon the vessel being maintained, equipped and operated in accordance with the documented arrangements contained in the Compliance Examination and Declaration report form SCV2. Proposals to change any of the arrangements should therefore be agreed in writing with the Certifying Authority before a change is implemented. Copies of the written agreement detailing change(s) should be appended to the report form SCV2, which is to be retained on board the vessel.

27.9.2.2 When the vessel is found not to have been maintained or equipped or operated in accordance with the arrangements documented in the Compliance Examination and Declaration report form SCV2, the certificate may be cancelled by the Certifying Authority which issued the certificate.

27.9.2.3 When a vessel is sold, the certificate issued by the Certifying Authority on the basis of the compliance examination and owner's declarations documented in the Compliance Examination and Declaration report form SCV2 is cancelled automatically and the owner/managing agent should return the certificate to the Certifying Authority for formal cancellation and records.

27.9.2.4 When a vessel has had its certificate cancelled, the Certifying Authority should report the circumstances to the Department of Transport for action to be taken as deemed necessary.

28 Vessels Operating under Race Rules

28.1 A vessel chartered or operated commercially for the purpose of racing need not comply with the provisions of the Code provided that, when racing, it is racing under the rules of the Union Internationale Motonautique and the affiliated national authority in the country where the race is taking part.

28.2 The relief from compliance with the provisions of the Code which is permitted by 28.1 does not apply to a vessel taking part in an event created and organised with an intent to avoid the provisions of the Code.

29 Clean Seas

A vessel complying with the Code should meet international, national, regional and local requirements for the prevention of marine pollution which are applicable to the area in which the vessel is operating.

Responsibility for the vessel to be properly equipped and maintained to meet the requirements prevailing rests with the owner/managing agent.

It is also the responsibility of the owner/managing agent to ensure that a charterer of a vessel receives up-to-date and adequate information on prevention of pollution in the area in which the charterer intends to operate. The information may include the need to seek advice from local authorities, for which contact "points" should be given.

Requirements for preventing pollution of the sea:-

Sewage When the direct overboard discharge from a water closet is prohibited by administrations/ authorities in an area of operation, the provision of "holding tanks" of sufficient capacity to store waste for discharge to shore facilities may be needed for a vessel to comply.

Garbage The disposal of garbage into the sea is prohibited by the Merchant Shipping (Prevention of Pollution by Garbage) Regulations 1988, SI 1988 No.2292. Arrangements for the retention of garbage on board and for discharge to shore facilities should be provided. Arrangements should be varied as necessary to comply with special requirements which may be applied by administrations/authorities in the area in which a vessel operates. Reference should be made to Merchant Shipping Notice No. M.1389.

Oil Merchant Shipping Notice No. M.1240 which is to be read in conjunction with the Merchant Shipping (Prevention of Oil Pollution) Regulations 1983, SI 1983 No.1398, as amended, explains the extent to which a vessel operating in accordance with the Code should comply with the Regulations. Examples of simple oily-water separating arrangements which may be acceptable for small vessels are described in M.1240.

Guidelines for systems for handling oily wastes in machinery spaces of ships (including yachts) are provided in Merchant Shipping Notice No. M.1451. The guidelines apply to ships of which the keels were laid on or after 1 January 1992.

Means to prevent pollution by oil should be acceptable to administrations/authorities in the area in which a vessel operates.

ANNEX 1

DEPARTMENT OF TRANSPORT MERCHANT SHIPPING NOTICE NO. M.1194

THE STATUS OF PERSONS CARRIED ON UNITED KINGDOM SHIPS

This Notice is addressed to Shipowners, Charterers, Masters and Persons in charge of United Kingdom Ships

1. During an appeal case [1] heard in the High Court in 1983, the legal status of persons on board a United Kingdom ship came under close scrutiny; in particular the distinction between "persons engaged on the business of the vessel" and "passengers". As a result of the judgement made in this case it has been decided to give the following guidance regarding the status of persons when carried on board United Kingdom ships.

2. The current legal definition of a passenger is given in Section 26 of the Merchant Shipping Act 1949 which states:

   (1) In Part II of the principal Act (ie, the Merchant Shipping Act 1894), in the Merchant Shipping (Safety and Load Lines Conventions) Act. 1932, and in this Act, the expression 'passenger' means any person carried in a ship, except

   (a) a person employed or engaged in any capacity on board the vessel on the business of the vessel;
   (b) a person on board the vessel either in pursuance of the obligation laid upon the master to carry shipwrecked, distressed or other persons, or by reason of any circumstance that neither the master nor the owner nor the charterer (if any) could have prevented or forestalled; and
   (c) a child under one year of age.

   (2) In the Merchant Shipping (Safety and Load Lines Convention), Act 1932, and in this Act, the expression 'passenger steamer' means a steamer carrying more than twelve passengers. (This definition of a passenger steamer was subsequently amended by Section 17(2) of the Merchant Shipping Act 1964).

3. After carefully studying the Court's judgement of the case it is the Department's view that the only persons who can be considered as being lawfully 'employed or engaged on the business of the vessel' are those over the minimum school leaving age (about 16 years) who:

   (i) have a contractually binding agreement to serve on the vessel in some defined capacity and which could include carrying out such duties under training, or are
   (ii) duly signed on members of the crew.

4. In addition to noting the foregoing, it is recommended that whenever the carriage of passengers is contemplated on any vessel the contents of Merchant Shipping Notice No. 913 should be carefully studied.

Department of Transport
Marine Directorate
London WC1V 6LP
October 1985

---

[1] The appeal case referred to in this Notice was: Secretary of State for Trade v. Charles Hector Booth (master of the yawl "Biche") [1984] I All E.R.; [1984] L1.L.Rep p 26.

ANNEX 2

LIFE-SAVING APPLIANCES

Area of operation category	4	3	2	1	0
m= nautical miles	< 20m daylight & favour-able weather	< 20m	≥ 20m & < 60m	≥ 60m & < 150m	≥ 150m
Liferafts Note 1	Yes	Yes	Yes	Yes	Yes
Lifebuoys Note 3	< 15 pers 2 ; ≥ 15 pers 4	< 15 pers 2 ; ≥ 15 pers 4	< 15 pers 2 ; ≥ 15 pers 4	< 15 pers 2 ; ≥ 15 pers 4	< 15 pers 2 ; ≥ 15 pers 4
Lifebuoy - light & drouges	2 Note 3.2	2	2	2	2
Lifebuoy - Dan-buoy Note 2	None	None	None	None	None
Buoyant line Note 3.1	1 or 2	1 or 2	1 or 2	1 or 2	1 or 2
Lifejacket Note 4	100% Note 4.2	100%	100%	100%	100%
Parachute flares	0	4	4	6	12
Red hand flares	2	6	6	6	6
Smoke Signals	2 buoyant or hand held	2 buoyant or hand held	2 buoyant or hand held	2 buoyant	2 buoyant
Thermal protective Aids (TPA) Note 5	100%	100%	100%	100%	100%

LIFE-SAVING APPLIANCES

Area of operation category	4	3	2	1	0
Portable VHF	1 Note 6	1	1	1	1
406MHz EPIRB Note 7	None	None	None	1 MPT 1278 or 1259	1 MPT 1278 or 1259
SART	None	None	None	1 Note 8	1 Note 8
General Alarm ≥15 persons	None	None	Yes	Yes	Yes
Life-Saving Signals Table 2 x SOLAS No. 2 or 1 x SOLAS No. 1	Yes	Yes	Yes	Yes	Yes
Line Throwing Appliance	None	None	None	None	None
Training manual	Yes	Yes	Yes	Yes	Yes
Instructions for on board maintenance	None	None Note 9	None Note 9	Yes	Yes

Notes:-

1 Liferafts

1.1 Category 0 vessels should be provided with liferafts of such number and capacity that, in the event of any one liferaft being lost or rendered unserviceable, there is sufficient capacity remaining for all on board.

1.2 Category 1 vessels carrying 15 or more persons should carry liferafts in accordance with 1.1.

1.3 Liferafts on vessels identified in 1.1 and 1.2 should be of DTp approved type equipped with "SOLAS A PACK" and contained in grp containers. The liferafts should be stowed on the weather deck or in an open space and should be fitted with float free arrangements (hydrostatic release units) so that the liferafts float free and inflate automatically.

1.4 Category 1 vessels carrying 14 or less persons and vessels of categories 2, 3 and 4 should be provided with liferaft capacity to accommodate at least the total number of persons on board.

1.5 Liferafts on vessels identified in 1.4 should be of either DTp approved type or Offshore Racing Council (ORC) type. Liferafts should be equipped with "SOLAS B PACK" or, if ORC standard liferafts are fitted each liferaft should be provided with a "grab bag" containing the following equipment:-

.1 second sea anchor and line;
.2 a first aid kit;
.3 one daylight signalling mirror;
.4 one signalling whistle;
.5 one radar reflector;
.6 two red rocket parachute flares;
.7 three red hand flares;
.8 one buoyant smoke signal;
.9 one thermal protective aid for each person on board; and
.10 one copy of the illustrated table of life-saving signals (SOLAS No.2).

NOTE! To facilitate rapid abandonment in an emergency a 'grab bag' should be provided in a position accessible and known to all on board.

1.6 Liferafts on vessels identified in 1.4 may be either

.1 in approved grp containers stowed on the weather deck or in an open space and fitted with float free arrangements so that the liferafts float free and inflate automatically; or

.2 in grp containers or valise stowed in readily accessible and dedicated weathertight lockers opening directly to the weather deck.

2 Dan-buoy

A Dan-buoy is not required to be provided on a motor vessel.

3 Lifebuoys

3.1 On vessels carrying 15 or more persons, buoyant lines of not less than 18 metres in length should be attached to each of the two lifebuoys not fitted with a light and a drogue.

3.2 Lifebuoys on category 4 vessels need not be provided with lights.

4 Lifejackets

4.1 If the lifejackets are inflatable an additional 10% or 2, whichever is the greater, should be provided.

4.2 A sufficient number of lifejackets should be provided for children carried on the vessel.

4.3 Lifejackets on category 4 vessels need not be provided with lights.

5 Thermal Protective Aids

TPAs may be stowed in the 'grab bag' (see note 1.5).

6 Portable VHF

If a fixed VHF is fitted in a category 4 vessel a portable VHF need not be provided.

7 406MHz EPIRB

7.1 The 406MHz EPIRB should be installed in an easily accessible position ready to be manually released, capable of being placed in a liferaft, and capable of floating free and automatic activation if the ship sinks.

7.2 On ships of less than 15 metres in length and carrying 14 or less persons the 406MHz EPIRB may be stowed in an accessible place and be capable of being placed readily in a liferaft without being capable of floating free.

8 SART

A SART is not required if the 406MHz EPIRB provided has a 121.5MHz frequency transmitting capability and is of the non-float free type for placing in a liferaft.

9 Instructions for on board maintenance

Vessels operating on bare boat charter should be provided with instructions.

ANNEX 3

OPEN FLAME GAS INSTALLATIONS

1 General Information

1.1 Possible dangers arising from the use of liquid petroleum gas (LPG) open flame appliances in the marine environment include fire , explosion and asphyxiation due to leakage of gas from the installation.

1.2 Consequently, the siting of gas consuming appliances and storage containers and the provision of adequate ventilation to spaces containing them is most important.

1.3 It is dangerous to sleep in spaces where gas-consuming open-flame appliances are left burning, because of the risk of carbon monoxide poisoning.

1.4 LPG is heavier than air and, if released, may travel some distance whilst seeking the lowest part of a space. Therefore, it is possible for gas to accumulate in relatively inaccessible areas, such as bilges, and diffuse to form an explosive mixture with air, as in the case of petrol vapour.

1.5 A frequent cause of accidents involving LPG installations is the use of unsuitable fittings and improvised "temporary" repairs.

2 Stowage of gas containers

2.1 Gas containers should be stowed on the open deck or in a gas-tight enclosure opening on to the deck, so that any gas which may leak can disperse overboard.

2.2 Stowage should be such that containers are positively secured against movement in any foreseeable event.

2.3 In multiple container installations, a non-return valve should be placed in the supply line near to the stop valve on each container. If a change-over device is used, it should be provided with non-return valves to isolate any depleted container.

2.4 When more than one container can supply a system, the system should not be used with a container removed.

2.5 Containers not in use or not being fitted into an installation should have the protecting cap in place over the container valve.

3 Fittings and pipework

3.1 Solid drawn copper alloy or stainless steel tube with appropriate compression or screwed fittings are recommended for general use for pipework in LPG installations.

3.2 Aluminium or steel tubing or any materials having a low melting point, such as rubber or plastic, should NOT BE USED.

3.3 Lengths of flexible piping (if required for flexible connections) should be kept as short as possible and be protected from inadvertent damage. Also, the piping should conform to an appropriate standard.

4 Open-flame heaters and gas refrigerators

4.1 When such appliances are installed, they should be well secured so as to avoid movement and, preferably, be of a type where the gas flames are isolated in a totally enclosed shield where the air supply and combustion gas outlets are piped to open air.

4.2 In refrigerators when the burners are fitted with flame arrestor gauzes, shielding of the flame may be an optional feature.

4.3 Refrigerators should be fitted with a flame failure device.

4.4 Flueless heaters should be selected only if fitted with atmosphere-sensitive cut-off devices to shut off the gas supply at a carbon dioxide concentration of not more than 1.5 per cent by volume.

4.5 Heaters of a catalytic type should not be used.

5 Flame failure devices

A gas consuming device should be fitted, where practicable, with an automatic gas shut-off device which operates in the event of flame failure.

6 Gas detection

6.1 Suitable means for detecting the leakage of gas should be provided in a compartment containing a gas-consuming appliance or in any adjoining space or compartment into which the gas (more dense than air) may seep.

6.2 Gas detectors should be securely fixed in the lower part of the compartment in the vicinity of the gas-consuming appliance and other space(s) into which gas may seep.

6.3 A gas detector should, preferably, be of a type which will be actuated promptly and automatically by the presence of a gas concentration in air of not greater than 0.5 per cent (representing approximately 25 per cent of the lower explosive limit) and should incorporate an audible and a visible alarm.

6.4 When electrical detection equipment is fitted, it should be certified as being flame-proof or intrinsically safe for the gas being used.

6.5 In all cases, the arrangements should be such that the detection system can be tested frequently whilst the vessel is in service.

7 Emergency action

7.1 A suitable notice, detailing the action to be taken when an alarm is given by the gas detection system, should be displayed prominently in the vessel.

7.2 The information given should include the following:-

.1 The need to be ever alert for gas leakage; and

.2 When leakage is detected or suspected, all gas-consuming appliances should be shut off at the main supply from the container(s) and NO SMOKING should be permitted until it is safe to do so.

.3 NAKED LIGHTS SHOULD NEVER BE USED AS A MEANS OF LOCATING GAS LEAKS.

ANNEX 4

FIRE FIGHTING EQUIPMENT

LESS THAN 15 METRES IN LENGTH AND 14 OR LESS PERSONS (1)	LESS THAN 24 METRES IN LENGTH (OTHER THAN COL 1) (2)
One hand powered fire pump (outside engine) or one power driven fire pump (outside engine space)*, with sea and hose connections, capable of delivering one jet of water to any part of the ship through hose and nozzle. One fire hose of adequate length with 10mm nozzle and suitable spray nozzle; or One multi-purpose fire extinguisher to BS 5423 with minimum fire rating of 13A/113B or smaller extinguishers giving the equivalent fire rating (in addition to that required below). Fixed fire extinguishing in engine space which may consist of a portable fire extinguisher arranged to discharge into the space. Not less than one multi-purpose fire extinguisher to BS 5423 with minimum fire rating of 5A/34B provided at each exit from accommodation spaces to the open deck. In no case should there be less than two such extinguishers provided. At least two fire buckets with lanyards. Buckets may be of metal, plastic or canvas and should be suitable for their intended service. One fire blanket in galley or cooking area (BS 6575 - light duty type).	One hand fire pump (outside engine space or one pwer driven fire pump (outside engine space)*, with sea and hose connections, capable of delivering one jet of water to any part of the ship through hose and nozzle. One fire hose of adequate length with 10mm nozzle and suitable spray nozzle. Fixed fire extinguishing in engine space which may consist of a portable extinguisher arranged to discharge into the space. Not less than two multi-purpose fire extinguishers to BS 5423 with a minimum fire rating of 13A/113B. At least two fire buckets with lanyards. Buckets may be of metal, plastic or canvas and should be suitable for their intended service. One fire blanket in galley or cooking area (BS 6575 - light duty type).

* This may be one of the pumps required by section 11, when fitted with a suitable change over arrangement which is readily accessible.

ANNEX 5

ANCHORS AND CABLES

Loa + Lwl / 2	Anchor Mass		Anchor Cable Diameter			
	Main	Kedge	Main Chain	Main Rope	Kedge Chain	Kedge Rope
(metres)	(kg)	(kg)	(mm)	(mm)	(mm)	(mm)
6	8	4	6	12	6	10
7	9	4	8	12	6	10
8	10	5	8	12	6	10
9	11	5	8	12	6	10
10	13	6	8	12	6	10
11	15	7	8	12	6	10
12	18	9	8	14	8	12
13	21	10	10	14	8	12
14	24	12	10	14	8	12
15	27	13	10	-	8	12
16	30	15	10	-	8	12
17	34	17	10	-	8	14
18	38	19	10	-	8	14
19	42	21	12	-	10	14
20	47	23	12	-	10	14
21	52	26	12	-	10	14
22	57	28	12	-	10	16
23	62	31	12	-	10	16
24	68	34	12	-	10	16

Notes:-

1. Chain cable diameter given is for short link chain. Chain cable should be sized in accordance with EN 24 565:1989 (covering ISO 4565:1986 and covered by BS 7160:1990 - Anchor chains for small craft), or equivalent.

2. The rope diameter given is for nylon construction. When rope of another construction is proposed, the breaking load should be not less than that of the nylon rope specified in the table.

3. When anchors and cables are manufactured to imperial sizes, the metric equivalent of the anchor mass and the cable diameter should not be less than the table value.

ANNEX 6

MEDICAL STORES

A vessel operating in area category 2, 3 or 4 should carry medical stores as follows:-

Name of Item and Ordering Description	Quantity Required
FIRST AID KIT The following to be in a damp proof strong canvas bag, satchel or box with a strap for carrying: (1) 4 x triangular bandages with sides of about 90cm and a base of about 127cm. (2) 6 x standard dressings no 8 or 13 BPC (3) 2 x standard dressings no 9 or 14 BPC (4) 2 x extra large sterile unmedicated dressings 28cm x 17.7cm (5) 6 medium size safety pins, rustless (6) 20 assorted adhesive dressing strips medicated BPC (7) 2 sterile pads with attachments (8) 2 x packages each containing 15g sterile cotton wool (9) 5 pairs of large, disposable polythene gloves.	1*
PARACETAMOL 500mg tablets	50*
SEASICKNESS REMEDY Tablets (Hyoscine hydrobromide 0.3mg recommended)	50*
BUTTERFLY CLOSURES Adhesive skin closures, length about 5cm individually sealed sterile, in a container	20* & **
FORCEPS Epilation with oblique ends, 12.5cm of stainless steel throughout	1
SCISSORS About 18cm, one blade sharp pointed and the other round-ended; conforming to BSI standard BS3646, published on 19/07/63	1
THERMOMETER Ordinary range clinical thermometer, stubby bulb pattern	1**
FIRST AID MANUAL Published by St. John Ambulance/ St. Andrews Ambulance Association/British Red Cross Society (latest edition)	1

* Twice these quantities to be carried in vessels carrying 15 or more persons
** Not required in area category 4

ANNEX 7

THE MANNING OF SMALL VESSELS

THE MANNING OF SMALL MOTOR VESSELS IN COMMERCIAL USE

This Annex gives information relating to the manning and operation of small motor vessels in commercial use as follows:

Section 1 -	Areas of Application
Section 2 -	Minimum Qualifications of the person in charge of the vessel and the additional person when required to be carried
Section 3 -	DTp Boatman's Licences
Section 4 -	Revalidation of Certificates & Licences
Section 5 -	Approved Engine Course
Section 6 -	Responsibility of the Owner/Managing Agent for the Safe Manning of the vessel
Section 7 -	Keeping a Safe Navigational Watch
Section 8 -	Withdrawal of Certificate
Section 9 -	Phasing-in Arrangements

General

Vessels of less than 80 tons gross or under 24 metres in length carrying not more than 12 passengers, being commercially operated motor vessels as defined in section 1.4 of the Code, and which comply with the requirements of the Code will be exempt from the need to comply fully with the Merchant Shipping (Certification of Deck Officers) Regulations 1985 and the Merchant Shipping (Certification of Marine Engineer Officers and Licensing of Marine Engine Operators) Regulations 1986 provided the manning of the vessel, when operating in the areas described in 1 below, is in accordance with the standards given in 2 below.

1 Areas of Application

Commercially operated motor vessels operating within the following areas should carry at least the qualified personnel shown in 2 below:-

Area category 3 or 4	Up to 20 miles from a safe haven
Area category 2	Up to 60 miles from a safe haven
Area category 1	Up to 150 miles from a safe haven
Area category 0	Unrestricted service

2 Minimum Qualifications of the Person In Charge of the Vessel (Skipper) and of the Additional Persons Required to be Carried On Board

2.1 Endorsement of Certificates

All certificates of competency and/or service should carry the endorsement - "valid for pleasure vessels of up to 24 metres in length used for commercial purposes".

2.2 Qualifications Required

2.2.1 Voyages up to 20 miles from a safe haven - operating area category 3 or 4

The skipper should hold at least an RYA/DTp Certificate of Competency as Coastal Skipper (Motor).

2.2.2 Voyages of up to 60 miles from a safe haven - area category 2

The skipper should hold at least an RYA/DTp Certificate of Competency as Yachtmaster Offshore (Motor).

There should also be on board a second person deemed by the skipper to be experienced.

2.2.3 Voyages of up to 150 miles from a safe haven - area category 1

The skipper should hold at least an RYA/DTp Certificate of Competency as Yachtmaster Offshore (Motor).

There should also be on board a second person holding an RYA/DTp Certificate of Competency as Coastal Skipper (Motor).

One of the persons referred to above should be familiar with the operation and maintenance of the main propulsion machinery of the vessel, and should have attended an approved engine course.

2.2.4 Unrestricted Service - area category 0

The skipper should hold at least an RYA/DTP Certificate of Competency as Yachtmaster Ocean (Motor).

There should also be on board another person holding at least an RYA/DTp Certificate of Competency as either Yachtmaster Ocean or Yachtmaster Offshore (Motor).

One of the persons referred to above, or another person, should be familiar with the operation and maintenance of the main propulsion and associated machinery of the vessel and should have attended an approved engine course.

2.2.5 Radio Qualifications

Every vessel should carry at least one person holding a Radio Operator's Certificate suitable for the radio equipment on board.

2.2.6 Medical Fitness Certificates

The skipper should hold a Medical Fitness Certificate issued by the DTp or an equivalent certificate. A DTp Medical Report on an applicant for a Boatmaster's Licence or a Health and Safety Executive Medical report for a commercial sea diver will be considered to be equivalent to a DTp Medical Fitness Certificate.

2.2.7 Basic Sea Survival Course

Skippers of vessels to which the Code applies should hold an approved Basic Sea Survival Course Certificate.

2.2.8 First Aid Courses

Skippers or another member of the crew of vessels which operate in area category 2, 3 or 4 should hold a DTp First Aid at Sea Certificate or a certificate issued by a voluntary society following the successful completion of a first aid course approved by the Health and Safety Executive. Such courses should have extra emphasis on the treatment of hypothermia and casualty evacuation.

Skippers of vessels operating in area category 0 or 1 should hold a DTp Ship Captain's Medical Training Certificate unless another member of the crew holds a medical or nursing qualification of an equivalent or a higher standard.

3 DTp Boatman's Licences

Persons holding DTp Boatman's Licences issued prior to this Code of Practice coming into force should be considered, subject to them having the relevant experience, to be properly qualified only for the voyages in area category 2, 3 or 4, provided that only one nominated safe haven is indicated on the vessel's certificate.

4 Revalidation of Certificates and Licences

All RYA/DTp Yachtmaster Certificates, whether of competency or service, and existing Boatman's Licences should be revalidated every five years. To revalidate, the applicant should prove at least 150 days of actual sea service on motor vessels during the previous 5 years and be in possession of a valid Medical Fitness Certificate.

In order to revalidate a Boatman's Licence for use in vessels to which this section applies, the holder will have to produce evidence of having attended a Basic Sea Survival Course.

5 Approved Engine Course

An approved engine course is a shore based course of at least thirty hours duration which is approved or recognised by the Department of Transport. A "Certificate of Attendance" will be given by the course organisers to persons completing the course.

6 Responsibility of the Owner/Managing Agent for Safe Manning of the Vessel

It is the responsibility of the owner/managing agent to ensure that the skipper and where necessary the crew of the vessel have, in addition to any qualifications required in 2 above, recent and relevant experience of the type and size of vessel, the machinery on the vessel, and the type of operation in which the vessel is engaged. The owner/managing agent should also ensure that there are sufficient additional crew on board having regard to the type and duration of voyage being undertaken.

7 Keeping a Safe Navigational Watch

It is the responsibility of the skipper to ensure that there is, at all times, a person with adequate experience in charge of the navigational watch. In taking this decision the skipper should take into account all the factors affecting the safety of the boat, including:-

.1 the present and forecast state of the weather, visibility and sea;
.2 the proximity of navigational hazards;
.3 the density of traffic in the area.

8 Withdrawal of Certificate of Competency or Service

The Yachtmaster Qualifications Panel reserves the right to withdraw a RYA/DTp Certificate of Competency or Certificate of Service at any time if due cause is shown.

9 Phasing-in Arrangements

9.1 When the Code comes into operation and until 31 March 1996, existing skippers who do not already hold the Certificates of Competency required by the Code, will be eligible to be issued with a Certificate of Service appropriate to their previous experience.

The Certificates of Service will be issued by the RYA to the applicant upon satisfactory proof of sea service.

Applicants for Certificates of Service should also obtain the appropriate qualification in First Aid.

9.2 Applicants for RYA Certificate of Service (COS)

.1 Applicants for Coastal Skipper COS should have a total of at least two years experience of which at least 100 days should have been spent actually at sea. Included in this two years at least one year, which includes at least 50 days actually at sea, should have been served as skipper of a small commercial motor vessel.

.2 Applicants for Yachtmaster Offshore COS should have a total of at least five years experience of which at least 250 days should have been spent actually at sea. Included in this five years at least two years, which includes at least 100 days actually at sea, should have been served as skipper of a small commercial motor vessel. Additionally, the required sea service should include at least 12 voyages of over 60 miles and at least 6 of these voyages should have been served in the capacity of skipper.

ANNEX 8

HANDOVER PROCEDURES FOR OWNERS/MANAGING AGENTS WHEN BARE-BOAT CHARTERING A VESSEL

1 Familiarisation at Handover

1.1 The owner/managing agent or appointed representative with intimate knowledge of the vessel should be present at the handover of the vessel to the chartering skipper and crew in order to complete the following familiarisation process:-

.1 A demonstration of the stowage of all gear and the method of use of all lifesaving and firefighting appliances on board the vessel should be given;

.2 The location and method of operation of all sea cocks and bilge pumps should be explained;

.3 A demonstration to ensure familiarisation with all mechanical, electrical and electronic equipment should be carried out;

.4 Details of routine maintenance required for any equipment should be declared;

.5 Checks to be carried out on the engine prior to starting, whilst running and after stopping to be demonstrated;

.6 The method of setting, sheeting and reefing each sail should be shown.

2 Documentation

2.1 The owner/managing agent or appointed representative, as detailed in 1. above, should ensure that the Vessel's File is shown to the chartering skipper. The Vessel's File should contain at least the following:-

.1 Registration papers

.2 Copies of the insurance policy

.3 Other necessary certificates

.4 Details of permitted operating area

.5 Instruction manuals

.6 Electrical wiring and piping/plumbing diagrams

.7 Inventory of the vessel's equipment

.8 Plan(s) showing the stowage position of all the movable equipment necessary for the safe operation of the vessel

.9 A list of names and telephone numbers (both in and out of office hours) of persons who may be contacted if the chartering skipper or the vessel is in need of assistance.

2.2 The skipper chartering the vessel should sign an acceptance note after the handover procedure with regard to the inventory, condition of items demonstrated, and the amounts of fuel and other consumable items on board which may be chargeable.

3 Procedure on return of the vessel to the owner/managing agent

3.1 At the end of the charter the owner/managing agent or appointed representative together with the chartering skipper should be present and the following procedure observed:-

.1 The vessel should be inspected;

.2 The vessel's inventory should be checked;

.3 Any damage, defect, losses, or need for repair should be listed.

3.2 The above details should be noted on an appropriate form which is to be signed by the owner/managing agent or appointed representative and the chartering skipper.

ANNEX 9

SKIPPERED CHARTER - SAFETY BRIEFING

1 Before the commencement of any voyage the skipper should ensure that all persons on board are briefed on the stowage and use of personal safety equipment such as lifejackets, thermal protective aids and lifebuoys, and the procedures to be followed in cases of emergency.

2 In addition to the requirements of 1, the skipper should brief at least one other person who will be sailing on the voyage regarding the following:-

.1 Location of liferafts and the method of launching;

.2 Procedures for the recovery of a person from the sea;

.3 Location and use of pyrotechnics;

.4 Procedures and operation of radios carried on board;

.5 Location of navigation and other light switches;

.6 Location and use of firefighting equipment;

.7 Method of starting, stopping, and controlling the main engine; and

.8 Method of navigating to a suitable port of refuge.

Safety cards will be considered to be an acceptable way of providing the above information.

ANNEX 10

PHASE-IN TIMETABLE

EXISTING VESSELS

The table states the **YEAR of the phase-in date of 01 APRIL** for groups of vessels identified by bands of length overall and area of operation.

CATEGORY	4	3	2	1	0
AREA OF OPERATION	DISTANCE	FROM	SAFE	HAVEN	(miles)
LENGTH OVERALL (metres)	≤ 20 favourable weather; daylight	≤ 20	≤ 60	≤ 150	Unlimited
L<4.9	1999	1995	1995	1995	1995
4.9 ≤ L < 7.6	1997	1996	1996	1995	1995
7.6 ≤ L < 9.2	1998	1997	1996	1995	1995
9.2 ≤ L < 10.7	1998	1997	1997	1996	1995
10.7 ≤ L < 12.2	1999	1998	1997	1996	1995
12.2 ≤ L < 13.7	1999	1998	1997	1996	1995
13.7 ≤ L	1999	1998	1997	1996	1995

Notes:- At the request of the owner/managing agent, a vessel may be examined by the Certifying Authority for compliance with the Code and, if found to be in compliance, a certificate may be issued prior to the phase-in date. (27.5.2 refers to examinations completed within 6 months of the phase-in date.)

ANNEX 11

(Seal/Crest and Name of Issuing Authority)

SMALL COMMERCIAL VESSEL CERTIFICATE

Name of Vessel ....

Official No ....

Port of Registry ....

Gross Tonnage ....

Maximum No. of Persons
to be carried ....

Date of Build ....

Name & Address of Owner/Managing Agent...

....

....

....

Length Overall...

Load Line Length....................................

Unique Identification No.........................

This is to certify that the above named vessel was examined by

.... of ....

.... at .... on ....

and found to be in accordance with the requirements of the Code of Practice for the Construction, Machinery, Equipment, Stability and Examination of Motor Vessels, of up to 24 metres Load Line length, in commercial use and which do not carry cargo or more than 12 passengers, published by the Surveyor General's Organisation of the Department of Transport.

This certificate will remain valid until .... subject to the vessel, its machinery and equipment being efficiently maintained, annual examinations and manning complying with the Code of Practice, and to the following conditions:-

....

....

....

....

....

....

The permitted area of operation is:-

....

....

Issued at .... on .... 19 ....

For and behalf of ....

Name .... Signature ....

Date ....

(Details of Annual Examinations are recorded on the report form for Compliance Examination and Declaration, SCV2.)

Printed in the United Kingdom for HMSO

Dd 296412, C50, 7/93